First published 1994 by CRC Press
Taylor & Francis Group
6000 Broken Sound Parkway NW, Suite 300
Boca Raton, FL 33487-2742

Reissued 2018 by CRC Press

© 1994 by CRC Press, Inc.
CRC Press is an imprint of Taylor & Francis Group, an Informa business

No claim to original U.S. Government works

Library of Congress Cataloging-in-Publication Data

Mechanisms of Pesticide Movement into Ground Water
    editors, Richard C. Honeycutt, Daniel J. Schabacker.
        p.   cm.
    ". . .a compilation of research papers presented at the 1991
American Chemical Society Meeting in Atlanta, Georgia [o]n April 18,
1991"—Pref.
    Includes bibliographical references and index.
    ISBN 0-87371-926-3
    1. Pesticides—Environmental aspects—Congresses.   2. Ground water
pollution—Congresses.   3. Soil pollution—Congresses.
I. Honeycutt, Richard C., 1945-   . II. Schabacker, Daniel J.
TD427.P35M43   1994
363.73'846—dc20   93-47251

A Library of Congress record exists under LC control number: 93047251

Publisher's Note
The publisher has gone to great lengths to ensure the quality of this reprint but points out that some imperfections in the original copies may be apparent.

Disclaimer
The publisher has made every effort to trace copyright holders and welcomes correspondence from those they have been unable to contact.

ISBN 13: 978-1-315-89524-6 (hbk)
ISBN 13: 978-1-351-07434-6 (ebk)

Visit the Taylor & Francis Web site at http://www.taylorandfrancis.com and the
CRC Press Web site at http://www.crcpress.com

MECHANISMS of PESTICIDE MOVEMENT into GROUND WATER

Edited by
**Richard C. Honeycutt
and Daniel J. Schabacker**

CRC Press
Taylor & Francis Group
Boca Raton London New York

CRC Press is an imprint of the
Taylor & Francis Group, an **informa** business

Preface

In 1984, Cohen et al. reviewed leaching and monitoring data on 12 pesticides found in ground water in 18 states.[1] In 1986, Cohen et al. reported at least 17 pesticides being found in ground water in a total of 23 states.[2] Although the presence of pesticides in ground water has been observed, the mechanisms whereby pesticides move into ground water have not been well established. When pesticides are found in ground water of a farm, or around a heavily farmed area, one must ask the following questions: (1) Are the sources of pesticide input into the ground water *'point'* source or *'non-point'* source?; (2) What is the hydrogeology of the farm area impacted? Is the local area characteristically karst or sand?; and (3) How deep are the aquifers in the local area impacted?

On the surface, these questions may seem trivial to a casual observer, however, the questions and the answers to them become extremely important in the current pesticide regulatory atmosphere. For example, when considering how to label a pesticide for the best ground water protection potential, it would not be scientifically sound to label the pesticide not usable in a local area where the hydrogeologic conditions preclude movement of the pesticide into ground water. This could apply, for example, to an area where the aquifer is extremely deep (e.g. >200 ft), where there are no karst areas and where, although research has shown wells with detectable residues of pesticides, research has, however, shown the source of the pesticides in the wells to be a 'point' source. Conversely, the regulatory agency may want to require the labeling of a pesticide to prevent ground water contamination in a local area where the aquifer is shallow (<30 ft), the subsoil is sandy, and research has shown that in this local area, pesticide residues have been frequently found in ground water.

This book is a compilation of research papers presented at the 1991 American Chemical Society Meeting in Atlanta, Georgia on April 18, 1991. The purpose of the book is to disseminate information concerning the mechanism of movement of pesticides into ground water. The book is divided into four sections: (1) Background Information and Experimental Parameters; (2) Experimental Results; (3) Modeling the Movement of Pesticides into Ground Water; and (4) Industry, Farm, and Regulatory Perspectives Related to the Movement of Pesticides into Ground Water.

The first section, "Background Information and Experimental Parameters" by Triegel, et al., contains an overview of the fate of pesticides in the environment and develops a discussion of the water balance — runoff vs. leaching. Another paper, by Weber, describes background information on pesticide properties and behavior in soil. These two papers collectively provide the necessary background information for understanding the remaining three sections of the book.

The second section, "Experimental Results", examines data from three extensive experiments using practical methods to search for the mechanism of movement of pesticides into ground water. In one paper, Weber, et al. examine the mobility of pesticides in field lysimeters. The techniques described in this paper are on the leading edge of experimental research into the mechanism of movement of pesticides into ground water. Turco, et al. describe a more practical approach of comparing laboratory and field leaching data for pesticides. Bicki, et al. describe remediation techniques used to decontaminate soil at agrochemical facilities. DeMartinis et al. examines the natural and man-made modes of entry of pesticides into ground water in agronomic areas.

The third section of this book deals with theoretical modeling of pesticide movement from the soil surface into the aquifer. Dowling, et al. present data on the modeling of movement of methomyl in soil/water systems, and, in another paper in this section, Jones describes a model which predicts the actual degradation and movement of pesticides in ground water once the pesticide has entered the aquifer.

The last section of the book deals with industry, farm, and governmental perspectives related to the movement of pesticides into ground water. Jones describes recent advances in managing agricultural chemicals in ground water at the farm level. Gilding, of the National Agricultural Chemical Association, provides an overview of industry perspective on pesticide issues related to ground water. Briskin presents some enlightening information on the methods, results, and policy implications of the recent National Pesticide Survey undertaken by the EPA.

With regard to state regulatory information concerning pesticides in ground water, Leffingwell presents an assessment of leaching potential of pesticides as developed under the Pesticide Contamination and Prevention Act of California, one of the leading states in pesticide ground water research and regulatory implementation. The section on industry and governmental perspectives also deals with perspectives concerning pesticides in drinking water from a farm wife's perspective (Greiner). These papers will enlighten the reader about the more practical and realistic side of the issues facing the use of pesticides, and their presence, in ground water.

I would be somewhat remiss if I did not approach the issue of future pesticide-ground water research in this preface. Future plans must include an extension of the research so eloquently outlined in this book. The role of models will undoubtedly remain prominent in this effort, as will the continuing prospective and retrospective field experiments. Laboratory research, although extensively relied on in the past, should take a lower preference to field research since, as pointed out in this book, the movement of pesticides into ground water is an extremely complex event influenced not only by the local soil types, but by the local climatology and hydrogeology of the pesticide use area and, of course, the chemicophysical characteristics of the pesticides themselves. Industry, regulatory agencies, farm communities, and persons with environmental concerns over pesticide ground water issues will benefit from information gained from such an approach.

REFERENCES

1. Cohen, S. Z., Creeger, S. M., Carsel, R. F., Enfield, C. G., "Treatment and Disposal of Pesticide Wastes", Kreiger, R. F., Seiber, J. N., Eds., ACS Symposium Series No. 259, American Chemical Society: Washington, D. C., 1984, pp. 297–325.
2. Cohen, S. Z., Eiden, C., and Corber, M. N., "Monitoring Ground Water for Pesticides", ACS Symposium Series No. 315, American Chemical Society: Washington, D. C. 1986, pp. 170–196.

R. Honeycutt, Ph. D.

Richard Honeycutt, Ph.D., is the President of H.E.R.A.C., Inc. (Hazard Evaluation and Regulatory Affairs Company, Inc.). Dr. Honeycutt is an analytical biochemist and field research specialist/consultant engaged in exposure assessment of pesticides to humans and the environment. He is an active member of the Agrochemical Division of the American Chemical Society (ACS). He is also a member of the Committee on Environmental Improvement (CEI) of the ACS for which he serves as the Chairman of the Agrochemical Subcommittee. Dr. Honeycutt is also a member of the Pesticide Committee of the Interntional Commission of Occupational Health.

Dr. Honeycutt was born in Newport News, Virginia, and attended Anderson University (A.B. in Chemistry, 1967) and Purdue University (Ph.D. in Biochemistry, 1971). He was an NIH Predoctoral Fellow from 1969 to 1971 and a Postdoctoral fellow at the Smithsonian from 1972 to 1973. He has spent 20 years in the pesticide metabolism and environmental chemistry working at both Rohm and Haas Company and Ciba-Geigy Corporation before founding H.E.R.A.C., Inc. Dr. Honeycutt has published widely in the field of pesticide environmental chemistry and has edited six books in this area.

Daniel J. Schabacker, M.S., is an Environmentalist II at Ciba Plant Protection. Mr. Schabacker is an environmental fate chemist/contractor in the Ecochemistry Group of the Environmental Fate and Effects Department. He is an active member of the Agrochemical Division of the American Chemical Society (ACS) and serves on a number of committees. Mr. Schabacker is also a member of the Society of Quality Assurance (SQA).

Mr. Schabacker was born in Chatham, New Jersey, and attended Guilford College in Greensboro, North Carolina (B.S. Chemistry, 1984) and University of North Carolina at Greensboro (M.S. Chemistry - Organometallic Synthesis, 1991). Prior to graduate school, he worked at Union Carbide/Rhône-Poulenc Ag Company as a research assistant performing laboratory environmental fate studies. In addition to organometallic synthesis during his graduate school training, he was a research assistant in the UNCG Clothing and Textiles Department working with non-woven personal protective clothing as a substance barrier for agricultural workers. This involved both field and laboratory investigations. He has also worked for Quality Associates, Incorporated, a consulting firm which performs agrochemical laboratory and field FIFRA GLP quality assurance activities, and at Pharmacology Toxicology Research Laboratories as an Environmental Specialist performing field studies. Mr. Schabacker joined Ciba in May, 1992.

Contributors

Thomas J. Bicki, Department of Agronomy, University of Illinois, 1102 South Goodwin Avenue, Urbana, IL 61801.

Jeanne S. Briskin, U.S. Environmental Protection Agency, 401 M Street S.W., Washington, D.C. 20460

Sandra C. Cooper, Blasland, Bouck & Lee, 6800 Jericho Turnpike, Suite 210W, Syosset, NY 11791.

Ronald G. Costella, Department of Chemistry, USMA, West Point, NY 10996.

James M. DeMartinis, Blasland, Bouck & Lee, 6800 Jericho Turnpike, Suite 210W, Syosset, NY 11791.

Kathryn C. Dowling, Biological Test Center, McGaw, Inc., 2525 McGaw Ave., Irvine, CA 92713.

Alan S. Felsot, Food and Environmental Quality Lab, Washington State University, 100 Sprout Road, Richland, WA 99352.

Thomas J. Gilding, National Agricultural Chemicals Association, The Madison Bldg., Suite 900, 1155 15th St. NW, Washington, D.C. 20005.

Sandra Hayes Greiner, American Agri-Women, Route 2, Box 193, Keota, EA 52248.

Lei Guo, Triegel and Associates, Inc., 1235 Westlakes Dr., Suite 320, Berwyn, PA 19312.

Bruce R. Johnson, Environmental Protection Agency, Department of Pesticide Regulation, 1220 N St., P.O. Box 942871, Sacramento, CA 94271.

Russell L. Jones, Rhone-Poulenc Ag Company, P.O. Box 12014, Research Triangle Park, NC 27709.

Philip C. Kearney, Natural Resources Institute, Beltsville Agricultural Research Center, Agricultural Research Service, Room 208, Bldg. 003, Beltsville, Maryland 20705.

K. E. Keller, USDA-ARS, National Soil Tilth Laboratory, 2150 Pammell Dr., Ames, IA 50011.

E. J. Kladivko, Department of Agronomy, 1150 Lilly Hall, Purdue University, West Lafayette, IN 47907.

J. T. Leffingwell, Environmental Protection Agency, Department of Pesticide Regulation, 1220 N St., P.O. Box 942871, Sacramento, CA 94271.

Ann T. Lemley, Graduate Field of Environmental Toxicology, 202 MVR Hall, Cornell University, Ithaca, NY 14853.

Elly K. Triegel, Triegel and Associates, Inc., 1235 Westlakes Dr., Suite 320, Berwyn, PA 19312.

R. F. Turco, Indiana Water Resources Research Center and Department of Agronomy,1150 Lilly Hall, Purdue University, West Lafayette, IN 47907.

J. B. Weber, Crop Science Department, North Carolina State University, Raleigh, NC 27607.

M. Rose Wilkerson, Environmental Protection Agency, Department of Pesticide Regulation, 1220 N St., P.O. Box 942871, Sacramento, CA 94271.

Table of Contents

Introduction

Philip C. Kearney

The growing demand for food and fiber on a global basis has increased the use of water and pesticides in agriculture. Water withdrawals from ground water aquifers have increased steadily over the past 50 years, and are rapidly approaching one trillion gallons per day.[1] Approximately 66% of all ground water withdrawals is used for irrigation purposes in agriculture. Over the same duration pesticide use increased dramatically and reached a level of about 817 million pounds in U.S. agriculture in 1991.[2] It was the interaction of increased usage of these two substances, plus improved analytical capabilities, that resulted in the detection of ground water residues. These detections set into motion a new wave of legislation and research that is the basis of this volume on *Advances in Understanding the Mechanism of Movement of Pesticides into Groundwater*. The simultaneous detection of DBCP residues in ground water in California and aldicarb residues in well water on Long Island in 1979 triggered the current controversy over the safety of the nation's water supply.[3,4] The objective of this introduction is to examine the regulatory and research activities leading up to and resulting from increased interest focused on ground water research on pesticides.

Early pesticide regulatory documents addressed the water issue. Some of the first environmental guidelines were contained in PR Notice 70–15 issued in 1970.[5] These guidelines contain two brief statements calling for studies on pesticide leaching and movement of residues in surface waters. Subsequent regulations published by the U.S. Environmental Protection Agency in 1982 outlined a comprehensive analysis of all processes affecting pesticide behavior in the environment.[6] In the current volume, Johnson et al. discuss California's legislation enacted in 1985 that profoundly influenced that state's pesticide registration process. This legislation, the Pesticide Contamination Prevention Act (PCPA), required all registrants to supply chemical and environmental data to the California Department of Food and Agriculture so that "Specific Numerical Values" could be calculated. These numerical values were used to determine leaching potentials. Thus we have seen the evolution of regulations on leaching studies come from two brief statements in a 1970 document to 6100 studies generated by the 1985 PCPA in California used to develop a framework to control pesticide leaching.

In the first chapter Triegel and Guo present an overview of the behavior of pesticide in the soil environment, with a strong emphasis on the mathematical description of the factors affecting transport. While early environmental pesticide research focused on transport of pesticides through soils, it became apparent that resources would never be available to study the 600 plus registered pesticides in the more than 10,000 soil types across the United States. One of the early pioneers in pesticides mobility studies was Dr. Jerome Weber at North Carolina State University. Weber was one of the first scientists to attempt to rationalize the physical properties of soils and pesticides chemistry as a basis of predicting pesticide movement through soils. It is fitting that the second chapter in the current volume is authored by Weber who presents the accumulation of over two decades research and experimentation on the interrelationship of physical constants as a predictive tool for modeling movement. Modeling is expanded on in chapters by Jones, and by Dowling, Costella, and Lemley. Jones presents an overview on modeling rather than a critical discussion of each of the several available models. Particularly useful is the need to understand the differences between modeling in the unsaturated and saturated zones. The concept of linking the two systems is considered as well as an overall consideration of the choice of models depending on the desired decision. This awareness has resulted in concerted efforts by scientists to study more carefully the processes and pathways by which pesticides move through the vadous zone and into ground water. Jones then considers several examples of modeling the movement of aldicarb under different soil and climatic conditions. In contrast, Dowling et al. modeled a specific compound, methomyl, detected in ground water in a New York survey. These authors model the transport of methomyl under steady state conditions in a small laboratory column. The simulation studies provide a mechanism for controlling a number of important variables and the ability to impose a large number of experimental parameters not practical under field conditions. Specifically, soil organic carbon content was found to retard movement and accelerate degradation of methomyl in their system.

The advantages and problems associated with using field systems to study pesticide fate and movement are considered by Weber and Keller, and by Turco and Kladivko. Weber and Keller conducted extensive mobility studies in soil column field lysimeters. These systems, if used properly, have the great advantage of working with a relatively undisturbed soil core. Application of lime and plant cover were found to affect the leaching of metolachlor, primisulfuron and atrazine in these field lysimeters. The problem with lysimeters is that jacketing a large intact soil core or block can be a labor-intensive operation often requiring heavy field equipment. Turco and Kladivko provide an extensive treatise on the many factors affecting pesticide movement under field conditions. They deal in some depth with current theories of the importance of preferential flow as a mechanism for rapid transit of solutes in channeled systems. They also examine the pros and cons of using tile drain effluent for field scale measurements of pesticides mobility. Soil adjacent to tiles provides ideal systems that account for the contribution of natural channeling on pesticide movement. The installation of drain tiles, however, does initially disturb the soil over the tile which may obliterate existing channel systems.

Basic pesticide mobility studies have lead to practical recommendations for reducing leachate contamination on farm land. Jones has identified the two major sources of possible pollution, i.e., non-point and point source residues. Advances in research on timing, placement and amount of application hold promise for alleviating the non-source problems. Far more troublesome have been the problems associated with point source residues resulting from spills due to improper storage, handling, mixing, equipment, washing or disposal practices. Modern delivery systems are eliminating some of the loading and mixing problems experienced with the metal containers that require pouring the contents into the sprayer.

DeMartinis and Cooper expand on the problems associated with point source pollution from natural and man-made sources. Natural modes of entry include movement through natural fissures, cracks or channels that develop as the result of pedogenic processes or biological processes such as worm holes, and decayed root channels. There is considerable discussion among soil physicists at this time as to whether the early rapid transit of some pesticides by preferential flow pathways can be factored into the classical soil movement equations. The classification of preferential flow as a point source of entry may elicit some debate among theorists, but the argument is academic since movement through natural channels is an important mode of entry into subsurface aquifers. DeMartinis and Cooper have a good understanding of man-made wells and provide some common sense advice on keeping pesticides out of well water based on their experience in real world conditions.

Even though there are now better guidelines on preventing point source contamination by pesticides, there are still many sites where soils are heavily contaminated with pesticides from previous spills. Bicki and Felsot point out that there are an estimated 14,000 agricultural facilities in the United States that sell, mix, load or otherwise handle pesticides. Some of these sites now have very large soil residues from past operations. Remediation of these sites is now a subject of considerable debate. Biotechnology may offer some solutions in the future by providing engineered organisms specifically designed to metabolize concentrated soil residues. The molecular biologists are now facing the hard reality that not enough is known about microbial ecology to recommend a reliable bioremediation system with any guarantee of success. Mulbry and Seffens[7] have recently isolated and sequenced genes responsible for an early step in atrazine metabolism. Communication with Mulbry suggests that engineered organisms are not robust enough to successfully compete in the soil environment to effect any type of remediation at this time. Bicki and Felsot present a rather comprehensive overview of current remediation practices and discuss options in selecting a cleanup method. They describe in detail two large-scale landfarming remediation experiments from actual contamination sites. Landfarming is a dilution technique where the contaminated soil is mixed with fresh soil to enhance natural decomposition. The experiments were not entirely successful because timing of application is critical, i.e., it must be done prior to the onset of cold weather. Second, pesticides in soils from highly contaminated sites did not breakdown as rapidly as comparable sites receiving the same amounts of applied pesticide on uncontaminated soil. The latter situation suggest that the heavily contaminated soils may be biologically inactive from prolonged exposure to pesticides and other chemicals.

The National Pesticide Survey conducted by the U.S. Environmental Protection Agency is considered in some detail by Briskin. In many respects it is one of the most comprehensive soil surveys ever conducted on ground water for pesticide residues. A total of 126 pesticides and degradation products plus nitrates were monitored in a statistically designed study to represent the nation's 94,600 community water supply sources and 10.5 million rural domestic water supplies. The study was designed in 1985 and phase I was completed in 1990. The results showed that about 10.5% of the community and 4.2 percent of the rural water supply contained detectable levels of at least one pesticide. Fewer than 1% of the drinking water wells, however, contained one or more pesticides at levels of health concern. The most frequently detected product was dacthal, followed by atrazine, simazine, prometon, hexachlorobenzene, DBCP, dinoseb, ethylene dibromide and others. Most current ground water surveys are focusing on the preemergent, soil incorporated s-trizanes because of their extensive use in corn and other field crops.

Finally, ground water residues are discussed by different members of the agricultural community from their various perspectives. Industry's point of view is represented by Gilding, speaking for the National Agricultural Chemicals Association. Gilding argues that the National Pesticide Survey shows that the presence of pesticide residues in ground water is not a widespread occurrence, either in the number of residues detected, their frequency or concentration. Industry proposes that we identify the risk associated with each detection of a residue in ground water and manage that risk according to a specific location. The overall industry theme is that we must learn to use pesticides to maintain the nation's high level of productivity and learn to manage risks associated with low levels of ground water contamination.

REFERENCES

1. U.S. Geological Survey, National Water Summary 1987 — Hydrologic Events and Water Supply and Use. Water-Supply Paper 2350, Denver, CO, 1990.
2. Aspelin, A. L.; Grube, A. H.; and Torla, R. Pesticide Industrial Sales and Usage 1990–1991 Market Estimates, EPA Fall 1992.
3. Cohen, D. B. *Evaluation of Pesticides in Ground Water;* Garner, W. Y.; Honeycutt, R. C. and Nigg, H. N.; Eds.; ACS Symp. Ser. 315; American Chemical Society: Washington, D. C. pp. 499–529.

4. Zaki, M. H.; Moran, D.; and Harris, D. *Am. J. Public Health* 1982, 72. 1391–1395.

5. U.S. Department of Agriculture, PR Notice 70–15 on Guidelines for Studies to Determine the Impact of Pesticides on the Environment, June 1970.

6. U.S. Environmental Protection Agency, Office of Pesticides and Toxic Substances (1982) Pesticide Assessment Guidelines, Subdivision N, Chemistry: Environmental Fate. Washington, D.C.

7. Mulbry, W. and Seffens, W.; Characterization of the s-triazine hydrolase gene *trzA* from *Rhodococcus corallinus.* Abstract of the American Society for Microbiology.

CHAPTER 1

Overview of the Fate of Pesticides in the Environment, Water Balance; Runoff vs. Leaching

Elly K. Triegel and Lei Guo

The fate of pesticides in the environment is determined by many processes, which are functions of both the environmental conditions and the properties of the pesticides. All elements in the system, namely the water, the pesticides, and the soil, affect the transport processes. Water balance determines the potential leaching and runoff losses. However, the actual transport depends upon the soil characteristics, the field microenvironment, and the pesticide properties, such as water solubility and persistence. Those molecules which are not adsorbed and remain toxic present the principal concern for surface and ground water contamination. An integrated understanding of the transport mechanisms is essential for developing scientific strategies to deal with issues of environmental fate.

Utilization of pesticides for many years has been, and will continue to be, an important means to protect agricultural products. However, beneficial use of pesticides is not without attendant problems. Since 1979, at least 74 pesticides have been detected in ground water in 38 states.[1] There were 46 of these pesticide detections attributed to normal agricultural use and 32 were attributed to point sources. Contamination of ground water by pesticides has been of great concern in recent years since roughly half the U.S. population relies on ground water as a source of drinking water, and in rural areas this population may reach 90% or more.[2,3]

Most of pesticides are released to the environment either by agricultural application (non-point source or point source) or as a result of accidental spills or mishandling (point source). These compounds may eventually enter into soil, in which they may be subjected to processes such as volatilization, chemical

0-87371-926-3/94/$0.00+$.50

transformation, adsorption-desorption on soil constituents, and microbial degradation. Those pesticides surviving the chemical and microbial breakdown processes may move through the unsaturated zone and reach ground water. The rate and amount of pesticide transferred from the land surface to the ground water are governed by various factors, such as properties of the unsaturated/saturated zone, physicochemical nature of the chemical, and characteristics of the flow regime.[4-6] In addition, pesticides can directly reach ground water through artificial (e.g., drainage wells) or natural openings (e.g., macropores).

An integral understanding of the fate of pesticides in the environment will allow us (1) to monitor and/or assess the potential of ground water contamination from agricultural applications; (2) to design Best Management Practices that minimize potential impacts; and (3) to remedy the contamination caused by accidental pesticide spills.

WATER BALANCE

When the steady-state condition is considered, water flow through a given boundary can be calculated in terms of water balance. The equation of water balance (ΔM) is an account of all quantities of water input (I), water output (O), and water storage changes within a ground water system or a soil (ΔS) during a given period of time:

$$\Delta M = I + O \pm \Delta S$$

where I is the inflow, O is the outflow, and ΔS the change in storage. The items in the equation can include a number of factors as the following:

1. Inflow: precipitation, irrigation, and other imported water
2. Outflow: runoff, recharge to ground water, consumptive usage (including evapo-transpiration losses), and exported water
3. Storage: change in surface storage, change in ground water storage, change in soil moisture

For a given volume of soil, the water balance equation can be expressed as:

$$P + I + U = R + D + ET \pm \Delta S$$

where P is the precipitation; I is the irrigation inputs; U is the upward capillary flow into the soil zone; R is the runoff losses; D is the downward drainage (ground water recharge); ET is the evapotranspiration losses; and ΔS is the water storage during a given period of time. Each item is subject to change from time to time depending upon the other items.[7]

Regarding the migration of pesticides, the water balance of the system should be considered in terms of the temporal and spatial variations in climate and topography, antecedent moisture conditions (i.e., whether previous rainfall events have saturated the soil), and time of pesticide application relative to the rainfall event. Examples of temporal variations are

1. Warm, moist conditions during and after pesticide application would be conducive to biological degradation
2. A high intensity rainfall event immediately after application is more likely to produce pesticide loss through runoff
3. Comparatively dry conditions, followed by pesticide application, followed by moderate intensity rainfall events increase the probability of downward leaching
4. Hot dry weather following pesticide application would be more likely to result in volatilization than runoff or leaching

Factors affecting leaching and runoff are discussed more fully below.

LEACHING

Transportation of pesticides in soil may occur in many forms, including migration with water in the dissolved or suspended state, with soil particles in the adsorbed state, or with soil air in the vapor state. In terms of ground water contamination, however, pesticide movement by water flow is of the most practical significance. The movement of pesticide with water is governed by two processes: mass flow and dispersion.

Mass Flow

Pesticide movement by mass flow can occur in dissolved or suspended form. Mass flow happens as a response of water to a variety of forces, such as gravitational and capillary forces.

The Hagen-Poiseuille equation[8] showed that the flow rate of water through a straight circular tube is

$$\frac{\vec{V}}{\pi r^2} = -\frac{r^2 g}{8\gamma} \nabla \vec{H}$$

where V is the water flux, r the radius of the tube, g the gravitational acceleration, γ the viscosity of water, and ∇H the hydraulic gradient.

This equation shows that large pressures are required to force water through narrow pores and applies strictly only to laminar flow and Newtonian fluid conditions where no turbulent conditions happen at fast flow. Unfortunately, however, the pore system in soils is too complex to apply theoretical approaches developed for the analysis of laminar flow in such single tubes as those underlying Hagen-Poiseuille's equation. Nevertheless, Darcy, who studied rates of flow through columns of saturated sand, established a general law of the flow of homogeneous fluids in porous media which demonstrated that the rate of flow through a column was proportional to the potential difference between the ends of the column and inversely proportional to the distance between the ends:

$$\vec{V} = -K \cdot \nabla \vec{H}$$

where K is a constant called the hydraulic conductivity.[9]

Subsequent work[10] has shown that provided the moisture content is constant, Darcy's law holds for saturated as well as for unsaturated materials, since the air-filled pores act effectively as solids in their effects on the flow of the fluid. In light of that, the geometry of pore systems in natural materials is too complicated for the conductivity to be calculated theoretically from first principles; attempts have been made to relate conductivity to features such as pore size and porosity based on Hagen-Poiseuille's equation.[8]

The quantitative application of unsaturated flow theory to field or laboratory soil systems requires a knowledge of the relationship between soil hydraulic conductivity [$k\,(\theta)$] and soil-water characteristics (θ) relationships. Soil wetness at a given suction is generally greater in drying than in wetting. This dependence of the equilibrium content and state of soil water upon the direction of the process is called hysteresis.[11] In practice, flow theory is applied assuming there is a unique function that describes the soil-water characteristic, and effects of hysteresis are ignored.

Dispersion

In addition to mass flow, the movement of pesticides in porous media involves a distinguished process called "dispersion". In fact, dispersion process combines effects of molecular diffusion caused by a concentration gradient and mechanical mixing (hydrodynamic dispersion) caused by flow through a porous media. Molecular diffusion, which tends to make the concentration to be uniform throughout the system, occurs as a result of the thermal motions of the diffusing molecules. Hydrodynamic dispersion is attributed to the non-uniformity of velocity distributions that results from the characteristics of flow through narrow pores and from the complex geometry of the pore system.[12,13] Dispersion coefficient D_a can be expressed as:

$$D_a = \alpha v + \lambda D$$

where α is the dispersivity, v the flow velocity, λ the tortuosity coefficient, and D_a the coefficient of molecular diffusion.[14] Thus, the movement of the chemical in porous media by dispersion can be expressed as:

$$\frac{\partial C}{\partial t} = \frac{\partial^2}{\partial X^2} (D_a C)$$

Diffusion in porous media such as soil is slower than in free solution, because the pathway through the pores is restricted and tortuous, and the chemical interacts with the structural solids. According to the equation, the significance of molecular diffusion decreases as the flow velocity increases.[15]

RUNOFF

Pesticide losses from the original site by the force of surface water flow is often termed as runoff. Runoff occurs when input of water exceeds infiltration. Pesticide runoff includes losses from the dissolved and sediment-adsorbed pesticide. Though runoff generally results directly in the contamination of surface water, it can also contribute to ground water contamination through recharging ground water by the surface water.

Pesticides remaining in the top soil zone can be extracted into runoff through

1. Diffusion and turbulent transport of dissolved pesticide from soil pores to the runoff stream
2. Desorption from soil particles into the moving liquid boundary
3. Dissolution of stationary pesticide particulates
4. Scouring of pesticide particulates and their subsequent dissolution or suspension in the moving water[16]

The effective surface soil depth that interacts with runoff is around 2 to 10 mm.[17]

Factors affecting losses of runoff include climate, site topography, soil structure and texture, pesticide properties, and management. Rainfall timing with respect to pesticide application is particularly important; pesticides in the runoff-active zone decline with time rapidly by processes of volatilization, photolysis, leaching, and decomposition, generally following a first order pattern.[18,19] Pesticides with solubilities below 2 ppm are primarily lost through suspended soil particles.[17] Pesticides that are strongly adsorbed by soil will be retained at soil surface and especially susceptible to runoff. The impact of raindrops breaks soil aggregates, and suspends and dislodges soil particles. The erosion of water flow over soil surfaces is more significant than wind erosion.

Conservative tillage (reduced or no-tillage) may be the most promising best management practice for keeping soil erosion below tolerable levels.[20] It has been generally assumed that runoff of agricultural chemicals will be greatly reduced as conservative tillage systems are widely implanted. However, use of herbicides in such practices may be increased. The usefulness of any of the many kinds of tillage systems in reducing runoff or erosion can be fairly specific to a given site or chemical.[20,21]

PARTITION BETWEEN MULTIPHASES IN POROUS MEDIA

The most important aspect governing the fate of pesticides in soil-water systems is the nature of their partitioning between vapor, soil solution, and soil particles. Partitioning and solubility determine their mobility in the unsaturated/saturated zone. Partitioning of pesticides between soil phases depends on the physical-chemical properties of both the organic compound and soil medium of concern. The partition between solution and solid phases is determined by the solubility of the compound and the adsorption-desorption processes. That between solution or solid and gaseous phases is controlled by the volatilization process. Soil adsorptivity towards pesticides is the dominant factor retarding pesticides from leaching into ground water. Molecules not adsorbed by soil tend to move with water and may contaminate ground water.

Adsorption and Desorption

Adsorption is a process by which pesticide molecules transfer from mobile phases (liquid or vapor) to stationary phase (soil particles); desorption is the reverse process. Because adsorption-desorption reactions generally do not follow the single-valued relationship, hysteresis (i.e., the irreversibility of some adsorbed molecules) is often observed.[22] Adsorption is mostly described by using adsorption isotherms, where the amount adsorbed is plotted as a function of the equilibrium concentration of the adsorbing species in solution. Various mathematical relationships have been developed to express the relationship between the amount adsorbed and the equilibrium solution concentration. Langmuir and Freundlich equations are the commonly used adsorption isotherms.

The Langmuir adsorption equation can be shown by the following equation:

$$\theta = \frac{K'C_w M}{1 + K'C_w}$$

where θ is the concentration of occupied sites per unit of adsorbent; M is the concentration of adsorption sites; K' is an apparent equilibrium constant; and C_w is the equilibrium concentration of adsorbate in solution.[23]

The Freundlich equation is an empirical relationship between the amount adsorbed and the equilibrium solution concentration:

$$C_s = K_f C_w^{1/n}$$

where C_s is the amount adsorbed; C_w is the equilibrium concentration; K_f is a constant related to the amount of adsorption; and $1/n$ is a constant related to the type and degree of curvature of the isotherm.[23] The Freundlich equation has been used extensively to describe the adsorption of pesticides on soils. The degree of adsorption is often judged by comparing K_f values. But care must be taken when comparing K_f values, since the value of K_f can be greatly affected by both $1/n$ and the concentration range of C_s and C_w. When the value of n is unity, K_f becomes K_d, the adsorption/desorption coefficient or the soil/water distribution coefficient. K_d is an important parameter governing the chemical mobility. High adsorptivity generally indicates a low mobility.

The two most important characteristics determining soil adsorption towards pesticides are the organic matter content of the soil and the water solubility of the organic compound. Adsorption of nonionic organic compounds on soils and sediments depends often directly on the organic carbon content of the absorbing phase (e.g., Chiou et al.)[24] Koc, the adsorption constant on the organic carbon basis, is practically independent of the soil type. As adsorption measurements are generally tedious and sometimes difficult due to the low solubility of some pesticides in water, a considerable effort has been focused on indirect estimation techniques for predicting adsorption. Such techniques use physical-chemical properties of pesticides (e.g., water solubility, melting point, and octanol-water partition coefficient) to estimate their Koc values.[25] Recent studies showed that adsorption of organic molecules by soil-water systems is completed through a partitioning mechanism.[26]

Volatilization

In the unsaturated zone, pesticide molecules can migrate in the vapor state. Pesticide losses through volatilization can sometimes reach 80 to 90% of the total applied. The volatilization process may not directly affect ground water contamination, but can have profound effects on the environmental fate of pesticides.

The principal factors controlling the rate of volatilization are (1) the inherent vapor pressure of the pesticide of concern; (2) the distribution of the pesticide among the different phases in the unsaturated zone; and (3) the moisture status of the unsaturated zone.[27] Henry's law coefficient (H), the ratio of the vapor pressure of a compound to its water solubility at the same temperature, is a useful measure of volatilization tendency of dilute solutes from water. Cohen et al.[5] have indicated that most of ground water contaminants have a H of less than 2.1×10^{-3} atm-m^3/mol at ambient temperature, which makes these compounds less volatile in aqueous systems than benzene and carbon tetrachloride.

CHEMICAL TRANSFORMATION AND BIOLOGICAL DEGRADATION

Pesticides entering the soil and water are subjected to both abiotic and biotic transformations. Therefore, mobility alone is not a good indicator of ground water pollution potential. The combination of mobility, persistence, and a degradation pathway determines whether a compound will be degraded to an innocuous form during its residence in the unsaturated zone above the ground water.[28] Potential contamination of ground water from short-life compounds is much less than from persistent ones. It should be noted that more than one degradation pathway exists, and that daughter products may also impact ground water quality.

Chemical Transformation

Chemical transformations of pesticides can occur as solutes in the aqueous phase or in the adsorbed forms on solid phase surfaces. Naturally occurring chemical transformations in the field include hydrolysis, redox reactions, and other reactions such as photolysis. Hydrolysis and redox reactions are the most prevalent in the water and sediments. Hydrolysis refers to the cleavage of a bond of a pesticide molecule and the formation of a new bond with the oxygen atom of water thereafter. Abiotic oxidation of pesticides mostly occurs in surface waters, and abiotic reduction in bottom sediments and anaerobic waters.[29] Chemicals exposed to sunlight on the land surface can undergo direct phototransformations through an excited state as well as indirect phototransformations through reacting with other chemicals in an excited state.[30] The chemical transformation of hydrophobic and hydrophilic pesticides in an oxidizing environment is mainly through hydrolysis, acid, base and neutral substitution, elimination, and ring cleavage. These reactions are affected by pH, temperature, and ionic strength.[31]

Biological Degradation

Degradation through microbial activity is the primary pathway for pesticide detoxification in terrestrial and aquatic ecosystems. Microbial activity can lead to the complete breakdown of pesticide molecules. Basically, the microbial transformation of pesticides includes biodegradation, polymerization or conjugation, and accumulation. During biodegradation, microorganisms can either utilize pesticides as substrates for energy source and nutrients (mineralization) or cometabolize the chemicals by microbial metabolic reactions. The latter mechanism does not sustain growth of the responsible populations.[32] Polymerization or conjugation are reactions which join pesticide molecules with other pesticide or natural compounds.[33] Pesticides can also be accumulated into microorganisms by passive physical mechanisms.[34]

Interactions between pesticides and soil constituents such as adsorption and partition profoundly affect the biological activity. Cationic pesticides are ionically adsorbed to most clay minerals and soil organic matter. They are biologically available in the adsorbed state in most cases. Basic pesticides are ionically and physically adsorbed to some clays and organic soil colloids. Adsorption is greater and bioactivity is lower under neutral or alkaline conditions.[34] Other factors that affect microbial activities, such as temperature, moisture, nutrition content, and pH, also affect the biodegradation of pesticides. Under most circumstances, for example, decreasing soil moisture tends to decrease the rate of pesticide degradation. Soil pH is a much more complex influencing factor; it can affect not only the establishment of the microbial populations and their relative activities, but also the adsorption of ionic pesticides.[35]

Accelerated degradation, a phenomenon in which pesticide is more rapidly decomposed than normal, is sometimes observed when a pesticide is applied to the same site repeatedly. The build up and enhancement of microbial populations with improved capacity to degrade the pesticide, due to the incubation effects, can account for such an observation. Investigation of this effect can lead to development of methods of increasing bioremediation of pesticides in the cases of accidental spillages. The use of microorganisms and enzymes for treatment of pesticidal contamination may prove to be a feasible alternative to the existing physicochemical methods.[36,37]

Reaction Kinetics

There are two rate models for pesticide degradation: one is the "power rate model", the other is the "hyperbolic rate model".[38]

In the hyperbolic rate model, the rate depends on both the concentration of the pesticide and the sum of concentration and other terms. In the simplest case, these other terms are a single constant:

$$\frac{dC}{dt} = -\frac{k_1 C}{k_2 + C}$$

where k_1 is the constant representing the maximum rate approached with increasing concentration and k_2 the pseudoequilibrium constant depending on the concentration of a catalytic agent (i.e., the enzyme). The degradation rate (dC/dt) is a complex function of pH, moisture, temperature, etc.[32]

In the power rate model, the rate is proportional to some power of the concentration and can be expressed as:

$$\frac{dC}{dt} = -kC^n$$

where n is an apparent "order of the reaction", not necessarily an integer. Even though the power rate model is usually applied to reactions in homogeneous solution, it is also a special form of the hyperbolic model. Most of pesticide degradations follow the pseudo first order:

$$\frac{dC}{dt} = -kC$$

It is also observed that compounds at lower concentrations may persist for a much longer time than at higher concentrations. Alexander suggested the existence of thresholds was the principal cause for such a phenomenon.[32]

POINT VS. NON-POINT SOURCES

The point source of pesticide contamination can occur through artificial courses such as accidental spillage, improper handling of agricultural facilities, and haphazard release through drainage wells as well as natural courses such as macropores. Non-point sources are generally due to application of the pesticide over broad areas as part of the farming practices. Concerns have directed to releases in hydrogeologically sensitive regions or near man-made conduit systems. The risk assessment of the potential health and environmental risks posed by the pesticide contamination is needed before decisions on remediation processes can be made.

Modeling

Modeling movement of pesticide to ground water is the current approach to address ground water contamination problems either for point or non-point sources. Many research and screen models have been constructed, such as LEACHM (Leaching Estimation and Chemistry Model) by Wagenet and Hutson,[39] PRZM (the Pesticide Root Zone Model) by Carsel et al.,[40] and CREAMS (a field scale model for Chemicals, Runoff, and Erosion from Agricultural Management Systems) by Knisel et al.[41] To modify and estimate the parameters required in a specific model, the knowledge of the modeler with the chemicals of concern are important. Even with the limitations in accuracy, modeling can create a general picture of the expected spatial and temporal distribution of the chemical in question.[42] Modeling also provides guidance for the application of the remediation technologies such as biological treatment, chemical treatment, and landfarming. Contamination assessment is involved with data collection related to sources of the pesticide release, its type, quantities, and rates of disposal. Other types of data refer to the soil and ground water being subject to contamination.

Significance

The increasing refinement of analytical techniques has resulted in the ability to detect ever smaller quantities of pesticide in ground and surface water. Hence, the large number of cases of reported detections may be a function of greater analytical sensitivity, as well as the more common use and/or transport of the pesticide. The effects of increasing rare or spatially sparse events, as well as those events of small magnitude but more frequent occurrence, may now be detected even though the resultant concentrations are very low (e.g., 10 parts-per-trillion, reported by Bethem and Cornacchia[43]). Examples of low-impact or rare events include migration along man-made pathways (e.g., faulty well casing). Evaluation of the significance of these concentrations should consider both (1) the biological effects (whether very low concentrations are likely to produce impacts, and whether extrapolations to such low doses are accurate), and (2) the possibility that the concentrations are due to cross-contamination, and are not true indicators of water quality.

SUMMARY

The processes affecting the environmental fate of pesticides, and hence their movement into surface or ground water, include water-transport, adsorption-desorption on soil, volatilization, and chemical and microbial degradation. Leaching of pesticides into ground water by mass flow and dispersion occurs as a result of water infiltration. All elements in the system, namely the soil, the water, and the pesticide, affect the transport process. Soil properties can influence the behavior of both water and pesticide. The input of water (precipitation and irrigation) determines the potential of water flow. However, the actual water influx changes greatly over time and from soil to soil, depending mainly on the soil texture, structure, and field topography. Pesticide runoff can occur when the water input rate exceeds the infiltration rate.

Adsorption and persistence of the pesticide affect its concentration in ground water as well as in runoff; the former is a function of both pesticide and soil characteristics and the site microenvironment. A key factor determining pesticide attenuation by adsorption is the soil organic matter content, as is indicated by the relative consistence of Koc for a pesticide found in different soils. A high soil adsorptivity generally indicates a low mobility and, therefore, a low potential of contaminating ground water. The degradation aspects of a pesticide, combined with its water solubility, affect its potential of contaminating either surface or ground water.

REFERENCES

1. Ritter, W. F. *J. Environ. Sci. Health* 1990, *B25(1)*, 1–29.
2. Council for Agricultural Science and Technology. *Groundwater Quality;* Rep. 103 (ISSN b194-4088) CAST; Ames, IA, 1985.
3. National Research Council. *Pesticides and Groundwater Quality.* Issues and problems in four states; Natl. Acad. Press: Washington, DC, 1986.
4. Rothschild, E. R.; Manser, R. J.; and Anderson, M. P. *Ground Water* 1982, *20,* 437–445.
5. Cohen, S. Z.; Creeger, S. M.; Carsel, R. F.; and Enfield, C. G. In *Treatment and Disposal of Pesticide Wastes;* Krueger, R. F.; Seiber, J. N., Eds.; ACS Symp. Series No. 259; American Chemical Society: Washington, DC, 1984; pp. 279–325.
6. Perry, A. S.; Muszkat, L.; and Perry, R. Y. In *Toxic Organic Chemicals in Porous Media;* Gerstl, Z.; Chen, Y.; Mingelgrin, U.; Yaron, B., Eds.; Springer-Verlag: New York, 1990; pp. 16–34.
7. Bowen, R. *Ground Water;* Applied Science Publishers Ltd.: London, 1980.
8. Scheidigger, A. E. *The Physics of Flow through Porous Media;* University of Toronto Press: Toronto, Canada, 1960.
9. Darcy, H. *Les Fontaines Publique de la Ville de Dijon;* Dalmont, Paris, 1856.
10. Childs, E. C. and Collis-George, C. N. *Proc. R. Soc.* 1950, *201A,* 392–405.
11. Hillel, D. *Fundamental of Soil Physics;* Academic Press: New York, 1980.
12. Nielson, D. R. and Biggar, J. W. *Soil Sci. Soc. Am. Proc.* 1961, *25,* 1–5.
13. Biggar, J. W. and Nielson, D. R. *Soil Sci. Soc. Am. Proc.* 1962, *26,* 125–128.
14. Millington, R. J. and Quirk, J. P. *Trans. Faraday Soc.* 1961, *57,* 1200–1207.
15. Schwille, F. In *Pollutants in Porous Media;* Yaron, B.; Dagan, G.; and Goldshmid, J., Eds.; Springer-Verlag: New York, 1984; pp. 27–49.
16. Bailey, G. W.; Swank, A. R., Jr.; and Nicholson, H. P. *J. Environ. Qual.* 1974, *3,* 95–102.
17. Wauchope, R. D. *J. Environ. Qual.* 1978, *7,* 459–472.
18. Leonard, R. A. In *Environmental Chemistry of Herbicides;* Grover, R., Ed.; CRC Publ. Co.: Boca Raton, FL, 1988; pp. 45–89.
19. Leonard, R. A. In *Pesticides in the Soil Environment: Processes, Impacts, and Modeling;* Cheng, H. H., Ed.; Soil Sci. Soc. Am. Inc.: Madison, WI, 1990; pp. 303–350.
20. Felsot, A.; Mitchell, J. K.; and Kenemer, A. L. *J. Environ. Qual.* 1990, *19,* 539–545.
21. Sharpley, A. N. *Soil Sci. Soc. Am. J.* 1985, *49,* 1010–1015.
22. Bowman, B. T. and Sans, W. W. *J. Environ. Qual.* 1985, *14,* 270–273.
23. Bailey, G. W. and White, J. L. *Residue Review* 1970, *32,* 29–92.
24. Chiou, C. T.; Peters, L. J.; and Freed, V. H. *Science* 1979, *206,* 831–832.
25. Rao, P. S. C.; Nkedi-Kizza, P.; Davidson, J. M.; and Ou, L. T. In *Agricultural Nonpoint Sources and Pollution: Model Selection and Application;* Giorgini, A.; Zingales, F., Eds.; Elsevier Sci. Publ. Inc.: New York, NY, 1986; pp. 55–77.
26. Chiou, C. T. In *Toxic Organic Chemicals in Porous Media;* Gerstl, Z.; Chen, Y.; and Mingelgrin, U., Eds.; Springer-Verlag: New York, 1989; pp. 163–175.
27. Taylor, A. W. and W. F. Spencer. In *Pesticides in the Soil Environment: Processes, Impacts, and Modeling;* Cheng, H. H., Ed.; Soil Sci. Soc. Am. Inc.: Madison, WI, 1990; pp. 213–269.

28. Jury, W. A.; Focht, D. D.; and Farmer, W. J. *J. Environ. Qual.* 1987, *16*, 422–428.
29. Macalady, D. L.; Tratnyek, P. G.; and Grundy, T. J. *J. Contam. Hydrol.* 1986, *1*, 1–28.
30. Maecheterre, L.; Choudhry, G. G.; and Barrie, C. R. *Rev. Environ. Contam. Toxicol.* 1988, *103*, 61–126.
31. Wolfe, N. L.; Mingelgrin, U.; and Miller, G. C. In *Pesticides in the Soil Environment: Processes, Impacts, and Modeling;* Cheng, H. H., Ed.; Soil Sci. Soc. Am. Inc.: Madison, WI, 1990; pp. 103–168.
32. Alexander, M. *Environ. Sci. Technol.* 1985, *19*, 106–111.
33. Bollag, J.-M. and Loll, M. J. *Experientia* 1983, *39*, 1221–1231.
34. Bollag, J.-M. and Liu, S.-Y. In *Pesticides in the Soil Environment: Processes, Impacts, and Modeling;* Cheng, H. H., Ed.; Soil Sci. Soc. Am. Inc.: Madison, WI, 1990; pp. 169–211.
35. Weber, J. B. and Weed, S. B. In *Pesticides in Soil and Water;* Guenzi, W. D., Ed.; Soil Sci. Soc. Am. Inc.: Madison, WI, 1974; pp. 223–256.
36. Munnecke, D. M. *J. Agric. Food. Chem.* 1980, *28*, 105–111.
37. Finn, R. K. *Experientia* 1983, *39*, 1231–1236.
38. Goring, C. A. I. and Laskowski, D. A. In *Environmental Dynamics of Pesticides;* Haque, R. and Freed, V. H., Ed.; Plenum Press: New York, 1975; pp. 155–172.
39. Wagenet, R. J. and Hutson, J. L. *LEACHM: A Finite Difference Model for Simulating Water, Salt and Pesticide Movement in the Plant Root Zone;* Cotinuum Vol. 2. Version 2.0; New York State Water Resour. Inst. Cornell Univ.: Ithaca, NY, 1989.
40. Carsel, R. F.; Smith, C. N.; Mulkey, J. D.; Dean, L. A.; and Jowise, P. P. *User's Manual for the Pesticide Root Zone Model (PRZM): Release 1;* USEPA EPA-600/3-84-109; U.S. Government Print Office: Washington, DC, 1984.
41. Knisel, W. G., Ed.; *CREAMS: A Field-Scale Model for Chemicals, Runoff, and Erosion from Agricultural Management Systems;* USDA-SEA Conserv. Res. Rep. 26; U.S. Government Print Office: Washington, DC, 1980.
42. Yaron, B. *Agric. Ecosystems Environ.* 1989, *26*, 275–297.
43. Bethem, R.; Cornacchia, J.; and Segawa, R. In *Symposium on the Mechanisms of Movement of Pesticides into Ground Water;* Schabacker, D. J. and Honeycutt, R., Eds.; ACS Symp. 1991. (the current series).

CHAPTER 2

Properties and Behavior of
Pesticides in Soil

J. B. Weber

The behavior of pesticides in soil and water is regulated by the properties of the compounds and the media, and by climatic conditions. Important properties include ionizability (pK_A), water solubility (Ksp), vapor pressure (VP), soil retention (Koc), and longevity (T–1/2). Quaternary N pesticides, organic As and P acid pesticides, organometallic fungicides, dinitroaniline herbicides and growth regulators, metabolites of organophosphate insecticides, pyrethroid pesticides, and very low water soluble nonionic pesticides are strongly retained by soil colloids and relatively immobile in soils. The chemicals have half-lives that range from very short to very long in soils. It is their high retention by soil that keeps them from getting into ground water. Carboxylic acid herbicides and growth regulators, hydroxy acid pesticides, aminosulfonyl acid herbicides, amide and anilide herbicides, carbamate and carbanilate pesticides, fumigants, and highly water soluble nonionic pesticides are weakly retained by soil colloids and are relatively mobile in soils. The compounds have very short to moderate longevity in soils and this lessens their potential for getting into ground water. Basic pesticides, chlorinated hydrocarbon pesticides, phenylurea pesticides, and thiocarbamate pesticides are retained by soil in low to high amounts and have very short to long half-lives in soil. Koc vs. Ksp relationships differ for different chemical families, depending on the ionizing properties and the types of functional groups present. Soil pH regulates the amount of binding and the rate of degradation of ionizable compounds but normally only the rate of degradation of nonionic compounds. Longevity and soil reactivity of the chemicals determines whether or not the chemicals contaminate ground water.

The behavior of pesticides in the environment is regulated by the properties of both the compounds and the soil constituents, with which they come in contact, the hydrogeology of the area, and climatic factors.[1-3] Chemical properties of pesticides have previously been discussed according to classical chemical families such as carbamates, triazine, phenylureas, etc.[4-6] Properties such as chemical structure, molecular weight, melting point, and water solubility of pesticides were emphasized, but some discussion of ionizability, volatility, heat of solution, and lipophilicity were also included where the values were available. Several handbooks list and discuss additional properties such as molecular formula, physical state, decomposition temperature, solubility in various organic solvents, acute oral LD_{50} values, spectra data, specific gravity, and boiling point.[7-12] Other books contain some chemical property values but stress primarily preparation and mode of action of pesticides.[13-16] In previous publications pesticides were classified and discussed according to ionizability, water solubility, vapor pressure, parachor (empirical molecular volume), and planar molecular surface area.[17-20] More recently, Wauchope[21] and Wauchope et al.[22] published a pesticide data base of key properties, including soil retention (Koc) and longevity (T–1/2) values. A composite picture of all the chemical and physical properties of a pesticide would be the ideal circumstance for predicting the behavior of a given pesticide in the environment. However, in many cases these values are not all available and in most cases it is essential to know only key properties to provide reasonable answers. The key properties include ionizability, water solubility, volatility, soil retention, and longevity. Ionizability refers to the strength of ionizable functional groups present and whether the pesticide has basic, acidic, amphoteric, or nonionizable properties. The pK_A or negative log of the dissociation constant is an index of the acid or base strength of a compound and is defined as the pH at which half of the chemical is in the ionized form and half is in the nonionized form. Solubility refers to the solubility product (Ksp) for the parent molecule in deionized water at a specified temperature (usually 25°C) and pH. Volatility refers to the tendency of the parent molecule to enter the vapor state, and the vapor pressure (VP) in mm of Hg at 25°C is an index of this phenomenon. Soil retention (Koc) is an index of the binding capacity of the pesticide molecule to soil organic carbon, as determined from sorption studies. It was originally suggested as an index of the sorption capacity of nonionic organic chemicals by soils, and it was assumed that only the organic fraction of the soil adsorbed the chemicals. However, this is the case only for very lipophilic compounds. Polar organic molecules are readily bound to both organic and inorganic surfaces in soils.[17] The Koc value for a given compound is an average of the Kd value divided by the organic fraction of the soil for an assortment of soils, i.e., Koc = Kd/ Fraction of organic carbon. Koc values are commonly reported in pesticide data bases and in the literature for both nonionic and ionic compounds, but the values are at best only crude approximations of the retention of the chemicals by soils. Longevity of a pesticide is normally expressed in terms of half-life (T– 1/2) of the parent molecule, as determined under normal use conditions in the region where it is used.

The five key properties are utilized in discussing and comparing 18 chemical families in four categories of pesticides, strongly basic, basic, acidic, and

Table 1. Classification Scheme for Organic Pesticide Behavior in Soils

Category	Family	Table
Strongly basic	Quaternary N pesticides	3
Basic	Basic pesticides	4
Acidic	Carboxylic acid herbicides and growth regulators	5
	Hydroxy acid pesticides	6
	Aminosulfonyl acid herbicides	7
	Organic As and P acid pesticides	8
	Organometallic fungicides	9
Nonionic	Amide and anilide herbicides	10
	Carbamate and carbanilate pesticides	11
	Chlorinated hydrocarbon pesticides	12
	Dinitroaniline herbicides and growth regulators	13
	Fumigant pesticides	14
	Organophosphate insecticides	15
	Phenylurea pesticides	16
	Pyrethroid pesticides	17
	Thiocarbamate herbicides	18
	Misc., very low water soluble pesticides	19
	Misc., moderate to high water soluble pesticides	20
Summary table		21

nonionic, as listed in Table 1. The 18 families of pesticides include quaternary N pesticides, basic pesticides, five acidic pesticide families, and 11 nonionic pesticide families. Descriptive terminology for the ranges in values for the properties of the chemicals, which are generally logarithmic in nature, are included (Table 2).

QUARTERNARY N PESTICIDES

Quaternary N pesticides are very highly basic compounds (pK_A of 9 to 11), with very high water solubilities (Ksp of 1 to 100%) and very low potential for volatilization (VP of 0.01 to 0.15×10^{-6} mmHg at 25°C) (Tables 2 and 3). The compounds ionize completely in aqueous solutions to yield cationic species as

Table 2. Descriptive Terminology and Ranges for Chemical and Biological Properties of Organic Pesticides

Descriptive terminology[a]	Basicity (pK_A)	Acidity (pK_A)	Water solubility (Ksp) (mg/l)	Volatility (VP) (mmHg × 10^{-6})	Soil reactivity (Koc)	Longevity (T–1/2) (days)
Very low or very short	<2	>8	<10	<1	<10^2	<10
Low or short	2–4	6–8	10–10^2	1–10	10^2–10^3	10–30
Moderate	4–6	4–6	10^2–10^3	10–10^2	10^3–10^4	30–90
High or long	6–8	2–4	10^3–10^4	10^2–10^3	10^4–10^5	90–180
Very high or very long	>8	<2	>10^4	>10^3	>10^5	>180

[a] Adjectives describing the relative base or acid strength, water solubility, volatility, soil retention, or longevity.

Table 3. Properties of Strongly Basic Quaternary N Containing
         Pesticides[a]

Common name	pK$_A$	Ksp (%)	VP (mmHg × 10^{-6})	Koc	T–1/2 (days)
Difenzoquat	11	76	0.01	10^5	100
Diquat	11	70	0.01	10^6	500
Dodine	9	1.0	0.15	10^5	20
Mepiquat	11	100	0.07	10^5	500
Paraquat	11	70	0.01	10^6	500

[a] From the *Agrochemical Handbook*,[7] Wauchope,[21] Wauchope et al.,[22]
     Farm Chemicals Handbook,[9] *Herbicide Handbook*,[11] and manufactur-
     ers' technical data sheets; pK$_A$ = ionization constant, Ksp = water
     solubility, VP = vapor pressure, Koc = soil reactivity, T–1/2 = half-life.

shown for paraquat in Equation 1 where:

$$\text{Paraquat dichloride} \rightarrow \text{Paraquat}^{2+} + 2\,\text{Cl}^-$$

where :   Paraquat dichloride = chloride salt formulation of paraquat

$$\text{Paraquat}^{2+} = \text{divalent paraquat cation}$$

$$\text{Cl}^- = \text{chloride ion} \tag{1}$$

Quarternary N pesticides are readily sorbed to the cation exchange complex
of soils in exchange for inorganic cations.[17,23-25] Clay minerals are chiefly
responsible for binding quaternary N pesticides in mineral soils. The compounds are bound through coulombic forces reinforced by physical forces, and
soil Koc values are very high, ranging from 10^5 to 10^6 (Table 3). Plotted values
for the five chemicals are shown in the collection of points in the upper right
hand quadrant of Figure 1. Retention by the soil is not related to water
solubility of the chemicals due to their cationic nature and high retention by the
soil. Reported R$_f$ values for quaternary N pesticides in soil systems lie very
near to zero indicating that the chemicals are nearly immobile (Figure 2).

Longevity of the quaternary N pesticides in the environment ranges from
short to very long (T–1/2 = 20 to 500 days) (Tables 2 and 3). These compounds
have been reported to be degraded photochemically and microbiologically,[24]
but when sorbed on the internal surfaces of expanding-type clay minerals, such
as montmorillonite, degradation occurs at a very slow rate.[26]

BASIC PESTICIDES

The names and properties of 26 basic pesticides are listed in Table 4.
Chemical families represented include aniline, formamidine, imidazole, pyri-
midine, thiadiazole, triazine, and triazole. Basicity ranges from very low to
moderate (pK$_A$ = 1 to 5) (Tables 2 and 4). Basicity is dependent upon the

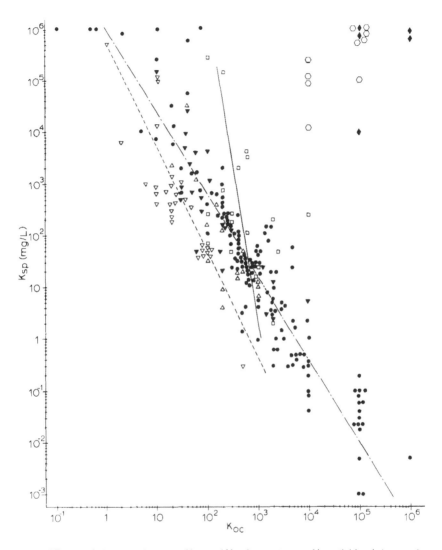

FIGURE 1 Relationships between Ksp and Koc for quaternary N pesticides (♦), organic As and P acid pesticides (○), basic pesticides with $pK_A > 2$ (□ and solid line), COOH acid herbicides (▽ and dashed line), and basic pesticides with $pK_A <$ 2 ( ), OH and $NHSO_2$ acid pesticides (▼), and nonionic pesticides (• and dash/dot line).

respective functional groups present, e.g., chloro-*s*-triazine compounds, such as anilazine, atrazine, cyanazine, propazine, and simazine, have very low basicity (Table 4, pK_A 1 to 1.8), while methoxy-*s*-triazines, such as prometon, and methylthio-*s*-triazines, such as ametryn, dipropetryn, prometryn, and terbutryn, have moderate basicity (Table 4, pK_A 4 to 4.3).[27]

Basic pesticides (B) associate with H+ ions in aqueous solutions and at acidic surfaces to form protonated species (HB+), as shown in Equation 2.

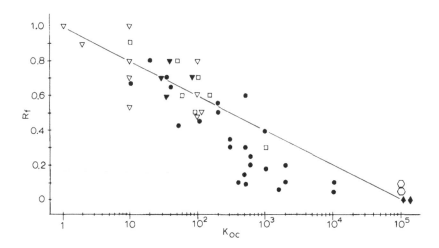

FIGURE 2   Relationship between R_f and Koc for quaternary N pesticides ($\blacklozenge$), organic As and P acid pesticides ($\bigcirc$), weakly basic pesticides ($\square$), COOH acid herbicides ($\triangledown$), OH and NHSO$_2$ acid pesticides ($\blacktriangledown$), and nonionic ($\bullet$) pesticides.

$$B + H^+ \leftrightarrow HB^+ \tag{2}$$

Since ionization is pH dependent, so too is water solubility (Ksp) and sorption on soil (Koc), as shown in Table 4. Basic pesticides range in water solubility from very low (Ksp = 2.0 mg/l) to very high (Ksp = 2.8×10^5 mg/l), and 24 (90%) of them have very low to low volatility (VP = 3×10^{-10} to 1×10^{-5} mmHg) (Tables 2 and 4). Soil retention of 20 (77%) of the basic pesticides is low (Koc = 100 to 1000) at neutral pH levels, but the values double or triple at pH $\leq$ of 5 (Table 4). Retention of three of the compounds is very low (Koc = 20 to 60) and retention of three others is moderate (Koc = 2000 to 10^4). Both soil organic matter and clay minerals have been reported to bind basic pesticides.[18,28-34] Although Ksp increases as pH decreases, Koc values increase because protonated species are readily sorbed to exchange sites in the soil (Table 4 and Figure 1). This is especially the case for low to moderately weak bases (pK$_A$ >2) where a specific relationship is plotted (solid line). For very weakly basic pesticides (pK$_A$ <2), the Ksp vs. Koc relationship generally follows the plotted line for nonionic pesticides. Reported Rf values for the weakly basic pesticides are generally inversely related to Koc values as shown in Figure 2.

Longevity of the basic pesticides ranges from very short to moderate (T–1/2 = 1 to 90 days) for 20 (77%) of the compounds and long to very long (T–1/2 = 100 to 360 days) for six others (Tables 2 and 4). For chloro-s-triazines, such as atrazine, and certain other pesticides where chemical hydrolysis is a major mode of degradation, degradation is very pH dependent and occurs faster under acidic conditions than under neutral or alkaline conditions.[28,35-37] For many

Table 4. Properties of Basic Pesticides[a]

Common name	pK$_A$	Ksp (mg/l)	VP (mmHg $\times 10^{-6}$)	Koc	T–1/2 (days)
Ametryn	4.10	185 pH 7 2000 pH 2	1.7	300 pH 7 400 pH 5	60
Amitrole	4.2	2.8×10^5	0.44	100	14
Anilazine	1	10	0.1	1000	1
Atrazine	1.68	33 pH 7 129 pH 2	0.6	100 pH 7 200 pH 5	60 pH 7 30 pH 5
Benomyl	5	2.0	0.037	2000	200
Chlordimeform	4	250	0.4	10^4	30
Cyanazine	1.0	160 pH 7 400 pH 2	0.02	100 pH 7 150 pH 5	20
Cyromazine	5	1.4×10^5	0.0003	200	100
DCNA	2	7.0	2.6	1000	10
Dipropetryn	4	16 pH 7 100 pH 2	1.6	900	30
Etridiazole	2	50	200	500	20
Fluridone	1.7	12	0.1	1000	21
Hexazinone	1.1	3.3×10^4	0.22	50	180
Isazophos	4	250	87	100	30
Metribuzin	1.0	1220	10	60	20
Penconazole	4	70	3.0	100	60
Prometon	4.28	750 pH 7 4168 pH 2	5.0	200 pH 7 600 pH 5	200
Prometryn	4.05	48 pH 7 3260 pH 2	2.0	300 pH 7 600 pH 5	60
Propazine	1.85	9 pH 7 19 pH 2	0.06	200 pH 7 400 pH 5	90 pH 6 60 pH 5
Propiconazole	4	110	1.0	600	60
Simazine	1.65	4 pH 7 16 pH 2	0.014	200 pH 7 400 pH 5	90 pH 7 60 pH 5
Tebuthiuron	1.2	2300	4.5	20	360
Tebutryn	4.3	25 pH 7 200 pH 2	2.1	1000 pH 7 2000 pH 5	40
Thiabendazole	4.7	50 pH 7 250 pH 3	0.0004	2500	200
Thiademefon	4	260	0.1	300	26
Thidiazuron	2	20	<0.01	500	80

[a] See footnote in Table 3.

other pesticides such as prometryn, where microbial degradation is a major mode of degradation, soil pH regulates the microbial populations and hence the rate of degradation.[36,38]

ACIDIC PESTICIDES

Five types of acidic pesticides are categorized and discussed according to their relative acidity, as determined by their ionizing groups (COOH, OH, or $NHSO_2$), presence of reactive groups (As or P) which readily interact with Fe, Al, and Ca in soils, or which are very insoluble organometallic salts. The carboxylic, hydroxy, and aminosulfonyl acid pesticides share some common properties and the As and P acid pesticides and organometallic pesticides share some common properties, but the groups must be discussed separately for an understanding of their behavior in soils. The first group (COOH, OH, NHSO$_2$ acid pesticides) are normally less retentive to soil colloids and more mobile in soils, and their sorption to soil is pH dependent. Reported R$_f$ values for some carboxylic, hydroxy, and amino sulfonyl acid pesticides show that mobility in the soil is generally inversely related to soil Koc values (Figure 2). The second group (As and P acid pesticides) are normally highly retained by soil colloids or form insoluble salts in soils and are nearly immobile. Reported R$_f$ values for As and P acid pesticides are very low and are inversely related to the very high Koc values reported for the compounds (Figure 2).

Carboxylic Acid Herbicides and Growth Regulators

Twenty-nine carboxylic acid herbicides and growth regulators are listed along with their properties in Table 5. Chemical families represented include aliphatic, benzoic, phenoxyalkanoic, phenylalkanoic, pyridinecarboxylic, pyridinyloxyalkanoic, quinoline carboxylic, and thiazolealkanoic acids. In all cases, the compounds are treated as ionizable acids even though some of them may be formulated as esters or soluble salts. In soil systems, ester formulations are rapidly hydrolyzed to free acid forms, and salt formulations equilibrate to free acid forms. Twenty-six (90%) of the chemicals have high to very high acidity (pK$_A$ = 0.7 to 4.0) and the remaining three compounds are moderately strong acids (pK$_A$ = 4.2 to 4.8) (Tables 2 and 5). (Second pK$_A$ values for a given compound will not be discussed.)

Carboxylic acid herbicides and growth regulators (HA) ionize in aqueous solutions to form anionic species (A–), as shown in Equation 3.

$$HA \leftrightarrow H^+ + A^- \qquad (3)$$

Ionization is pH dependent, as is Ksp and Koc (Table 5). With regards to water solubility, nine (31%) of the carboxylic acid compounds are low (Ksp = 50–100 mg/l), 12 (41%) are moderate (Ksp = 200–1000 mg/l) and eight (27%) are high to very high (Ksp = 1.25×10^4 to 1.2×10^6 mg/l) (Tables 2 and 5). The chemicals range in volatility from very low to moderate (VP = 10^{-8} to 33.7 × 10^{-6} mmHg).

Since soil K$_d$ and Koc values vary with pH, the pH of the media should always be reported along with the values. Unfortunately, however, this has not been a common practice when Kd and Koc values are reported in the literature. Koc values for carboxylic acid herbicides and growth regulators range from very low to low (Koc = 1 to 510) (Tables 2 and 5). Soil organic matter has been the major constituent in soil reported to bind acidic pesticides, but metallic hydrous oxides are also likely to be involved, particularly in tropical soils.[39-42] At

Table 5. Properties of Carboxylic (COOH) Acid Herbicides and Plant Growth Regulators[a]

Common name	pK$_A$	Ksp (mg/l)	VP (mmHg $\times$ 10^{-6})	Koc	T–1/2 (days)
Acifluorfen	2.2	50	<0.1	100	20
Bifenox	2.2	50	1.1	100	10
Chloramben	3.4	700	<0.1	15	14
Chlopyralid	2.3	1000	12	6	30
2,4-D	3.0	890 pH 6	8	20 pH 6	10
		500 pH 5		35 pH 5	
Dalapon	1.84	5 $\times$ 10^5	<10	1	20
Daminozide	4	1 $\times$ 10^5	<0.01	10	7
DCPA	3	50	<1	100	40
Dicamba	1.91	6500	33.7	2	14
Dichlorprop	2.86	350	<1	50	10
Diclofop	3.1	50	1.1	100	20
Endothall	3,6	1 $\times$ 10^5	<1	10	7
Fenac	3.9	203	<1	20	180
Fenoxaprop	3	50	0.01	100	9
Fluazifop	3	50	0.4	100	15
Imazamethabenz	2.9	1370 (p)	0.011	30	30
		857 (m)			
Imazapyr	1.8, 10.5	1.25 $\times$ 10^4	<0.01	10	90
Imazaquin	3.2, 10.2	600 pH 6	<0.01	30 pH 6	60
		60 pH 5		90 pH 5	
Imazethapyr	3.9, 10	1400 pH 6	<0.1	20	90
Lactofen	4	50	<1	100	3
MCPA	3.12	825	3.0	10	25
MCPB	4.8	660 pH 6	5	10 pH 6	21
		350 pH 5		20 pH 5	
NAA	4.2	420	<1	10	10
Naptalam	4.0	200	<1	20	14
Picloram	1.9	430	0.2	20	180
Quinclorac	4.8	62	<0.1	100	30
Quizalofop	3	100	<0.1	510	60
TCA	0.7	1.2 $\times$ 10^6	<2	1	20
Triclopyr	2.68	440	1.26	20	46

[a] See footnote in Table 3.

decreasing pH levels, Equation 3 is driven to the left, more molecular species are formed, and sorption, as indicated by Koc, is increased (Table 5). This is because the less soluble molecular species more readily bind to the lipophilic fraction of the soil organic matter complex than do the more soluble anionic species. It may also be in part due to less anionic species being present to be repelled by negatively charged soil colloids. Because of the low sorption of carboxylic acid compounds in soils, a specific Ksp vs. Koc relationship is plotted in Figure 1 (dashed line).

Longevity of 21 (72%) of the carboxylic acid herbicides and growth regulators ranges from very short to short (T–1/2 = 3 to 30 days), and 8 (28%) have moderate to long longevity (T–1/2 = 40 to 180 days) (Tables 2 and 5). Microbiological degradation by bacteria and actinomycetes, and to a lesser

Table 6. Properties of Hydroxy (OH) Acid Pesticides[a]

Common name	pK$_A$	Ksp (mg/l)	VP (mmHg × 10^{-6})	Koc	T–1/2 (days)
Bromacil	9.1	815	0.25	30	150
Bromoxynil	4.1	130	4.8	200	7
Chlorflurenol	7	21.8	49.9	200	2
Chlorobenzilate	7	13	0.9	2000	20
Dinoseb	4.5	52 pH 7	20	63 pH 7	20
		30 pH 4		500 pH 4	
DNOC	4.4	150	105	200	20
Ioxynil	4.0	50	7.5	200	7
MH	5.65	6000	<1	20	30
Sethoxydim	4.6	4700 pH 7	0.16	100 pH 7	5
		25 pH 4		600 pH 4	
Terbacil	9.0	710	0.31	50	120
Trichlorfon	6	1.5 × 10^5	15	10	10

[a]　See footnote in Table 3.

extent by fungi, is the major mode of degradation in soils.[39,43-47] Photodecomposition of the compounds also occurs, particularly in air and water, and when the compounds are present on leaf surfaces.[48]

Hydroxy Acid Pesticides

The properties of 11 hydroxy acid pesticides are listed in Table 6. Chemical families represented include hydroxycyclohexeneone, hydroxyfluorene, hydroxynitrile, hydroxyphosphonate, hydroxypyridinone, phenol, and uracil. Acidity of the chemicals ranges from moderate to very low (pK$_A$ = 4.0 to 9.1) (Table 2). Ionization in aqueous solutions occurs as depicted in Equation 3. Water solubility for the hydroxy acid pesticides is pH dependent, analogous to that of the carboxylic acid compounds, and ranges from low to very high (Ksp = 13.0 to 1.5 × 10^5 mg/l). Seven (64%) of the chemicals are very low to low in volatility (VP = 0.16 × 10^{-6} to 7.5 × 10^{-6} mmHg), three are moderately volatile (VP = 15 × 10^{-6} to 49.9 × 10^{-6} mmHg), and one is highly volatile (VP = 105 × 10^{-6} mmHg).

Soil retention of 10 (91%) of the hydroxy acid pesticides ranges from very low to low (Koc = 10 to 200) and retention of one is moderate (Koc = 2000) (Tables 2 and 6). Soil retention is pH dependent, analogous to that for the carboxylic acid compounds (Table 6). Soil organic matter has been the major soil constituent reported to bind hydroxy acid pesticides.[17,20,39,49] The relationship between Ksp and Koc for the hydroxy acid pesticides follows the same pattern as that for the nonionic pesticides (Figure 1).

With the exception of the uracils (bromacil and terbacil), longevity of the hydroxy acid pesticides is very short to short (T–1/2 = 2 to 30 days) (Tables 2 and 6). Bromacil and terbacil are somewhat persistent (T–1/2 = 120 and 150 days, respectively). Biological degradation is the principal mode of degradation of the hydroxy acid pesticides, with the exception of the uracils which are degraded chemically and biologically.[39,43,49,50]

Table 7. Properties of Aminosulfonyl (NHSO₂) Acid Pesticides[a]

Common name	pK_A	Ksp (mg/l)	VP (mmHg × 10⁻⁶)	Koc	T-1/2 (days)
Asulam	4.8	5000	<10	40	7
Bensulfuron	5.2	120 pH 7	<0.001	310 pH 7	5
		2.9 pH 5		2000 pH 5	
Bensulide	4	25 pH 7	0.8	1000 pH 7	90
		5.6 pH 5		10⁴ pH 5	
Bentazon	4	500	<0.01	30	20
Chlorimuron	4.2	1200 pH 7	<0.001	110 pH 7	30
		450 pH 6		150 pH 5	
		11 pH 5			
Chlorsulfuron	3.6	2.8 × 10⁴ pH 7	4.6	40 pH 7	90
		300 pH 5			
Fomesafen	3.0	4000 pH 6	21	60	30
		50 pH 4			
		2 pH 2			
Metsulfuron	3.3	9500 pH 6	<0.0001	35 pH 7	60
		1100 pH 5			
		270 pH 4			
Oryzalin	8.6	2.5	0.1	1500	20
Sulfometuron	5.2	300 pH 7	<0.0001	80 pH 7	30
Thifensulfuron	4.0	10 pH 5		50 pH 7	12
		2400 pH 6	<0.0001		
		260 pH 5			
		24 pH 4			

[a] See footnote in Table 3.

Aminosulfonyl Acid Herbicides

Table 7 contains the names and properties of 11 aminosulfony acid herbicides. Chemical families represented include benzothiadiazole, diphenylsulfone, sulfonylbenzamide, sulfonylcarbamate, sulfonyphosphorothioate, and sulfonylurea. With the exception of oryzalin, the chemicals range in acidity from moderate to high (pK_A = 3.0 to 5.2). Oryzalin is very low (pK_A = 8.6) in acidity. Ionization occurs as depicted in Equation 3. Water solubility for the aminosulfonyl acid herbicides is pH dependent, analogous to that for the carboxylic and hydroxy acid compounds. It ranges from high to very high (Ksp = 1200 to 2.8 × 10⁴ mg/l) for six (55%) of the compounds, moderate (Ksp = 120 to 500 mg/l) for three compounds, and very low (Ksp = 2.5 mg/l) for one compound. Ten (91%) of the compounds are low to very low in volatility and one (fomesafen) is moderately volatile.

Sorption of aminosulfonyl acid herbicides by soils is pH dependent, analogous to that for the carboxylic and hydroxy acid pesticides. Seven (64%) of the compounds have very low (Koc = 30 to 80) retention by soil, three have low (Koc = 110 to 1000) retention, and one has moderate (Koc = 1500) soil retention (Tables 2 and 7). Sorption of the aminosulfony acid herbicides in soils has been correlated with organic matter content, and mobility in soils has been inversely correlated with organic matter content.[51-53] The relationship between Ksp and Koc for the aminosulfonyl acid herbicides is similar to that for the hydroxy acid pesticides and nonionic pesticides (Figure 1).

Table 8. Properties of Organic Arsenic (As) and Phosphoric (P) Acid Pesticides[a]

Common name	pK$_A$	Ksp (mg/l)	VP (mmHg × 10^{-6})	Koc	T–1/2 (days)
Cacodylic acid	6.2	6.7 × 10^5	<0.1	10^5	42
DSMA	3.6, 8.2	7.0 × 10^5	<0.1	10^5	50
Ethephon	2,6	1 × 10^6	<0.1	10^5	10
Fosamine	2	1 × 10^5	4	10^5	8
Fosethyl	2	1.2 × 10^5	<0.1	10^4	1
Glyphosinate	2,6	2.5 × 10^5	<1	10^4	7
Glyphosate	2.3, 5.9, 10.9	1.2 × 10^4	<1	10^4	40
MAA	3.6, 8.2	1 × 10^5	<1	10^5	50
MAMA	3.6	1 × 10^6	<1	10^5	50
MSMA	3.6	1 × 10^6	<1	10^5	50

[a] See footnote in Table 3.

Longevity of the aminosulfonyl acid herbicides ranges from very short to moderate (T–1/2 = 5 to 90 days) (Tables 2 and 7). The chemicals are degraded in soil by both chemical and biological processes.[38,51] The sulfonylureas are more persistent under high pH regimes than under slightly acidic regimes where chemical hydrolysis proceeds more rapidly. Algae, fungi, actinomyces, and bacteria have been implicated in the biological degradation processes.

Organic As and P Acid Pesticides

The names and properties of ten As and P acid pesticides are given in Table 8. All of the compounds contain an ionizable As or P acid moiety. They ionize in aqueous solutions as depicted in Equation 3 and some have two or more pK$_A$'s. They range in acidity from high to low (pK$_A$ = 2 to 6.2) (Tables 2 and 8). The As and P acid pesticides are very highly water soluble (Ksp > 10^4) and very low to low in volatility (VP = <0.1 × 10^{-6} to 4 × 10^{-6} mmHg).

The As and P acid pesticides readily react with clay colloids and metallic hydrous oxides in soils and form insoluble precipitates with Fe, Al, and Ca.[54-58] They thus have high to very high soil retention (Koc = 10^4 to 10^5 (Tables 2 and 8).

The relationship between Ksp and Koc for the As and P acid pesticides is shown in the collection of points in the upper most corner of the upper right hand quadrant of Figure 1. Retention by the soil is not related to water solubility of the chemicals due to their high sorption by mineral colloids in soils and their readiness to form insoluble metallic precipitates.

Mobility (R$_f$) of the As and P acid pesticides in soils is very low and soil retention (Koc) very high, so reported values for the chemicals lie in the lower right hand quadrant of Figure 2.

Longevity of the As and P acid pesticides ranges from very short to short (T–1/2 = 1 to 50 days) (Tables 2 and 8). The P acid pesticides are degraded primarily by microbial degradation under both aerobic and anaerobic condi-

Table 9. Properties of Organometallic Fungicides[a]

Common name	Ksp (mg/l)	VP (mmHg × 10⁻⁶)	Koc	T–1/2 (days)
Mancozeb	6.0	<0.01	2000	70
Maneb	5.0	<0.01	2000	70
Metiram	0.1	<0.7	10^5	20
Triphenyltin hydroxide	0.2	0.007	10^4	75
Zineb	10	<0.01	2000	16
Ziram	65	0.1	1000	16

[a] See footnote in Table 3.

tions, and bacteria and fungi have been implicated in the degradation processes.[55,59] Degradation of the As acid pesticides is a multifaceted event involving cleavage of As from the organic carbon portion, degradation of the carbon portion, and reduction of As to a volatile form that escapes into the air.[34,60,61] Microbial degradation was the major pathway, and fungi, actinomycetes, and bacteria were implicated.

Organometallic Fungicides

All of the six organometallic fungicides listed in Table 9, except triphenyltin hydroxide, are low to very low water soluble Zn or Mn salts of dithiocarbamic acids (Ksp = 0.1 to 65 mg/l).[15] Triphenyltin hydroxide is a very low water soluble salt of Sn.[13] The compounds are all very low in volatility (VP = 0.007 × 10⁻⁶ to 0.1 × 10⁻⁶ mmHg).

The organometallic fungicides are strongly adsorbed by soil organic matter and/or exist as precipitated salts in soils.[14,62] Soil retention ranges from moderate to very high (Koc = 1000 to 10^5) (Tables 2 and 9). Because of the low Ksp values, the compounds lie in the lower right hand quadrant of the Ksp vs. Koc relationship and are included with the nonionic pesticides (Figure 1).

Longevity of the organometallic fungicides in soils ranges from short to moderate (T–1/2 = 16 to 75 days) (Tables 2 and 9). The compounds inhibit the growth of some actinomycetes and many fungi and are relatively noninhibitory to most bacteria.[63,64] They are degraded by photolytic, chemical, or biological mechanisms.

AMIDE AND ANILIDE HERBICIDES

The eight amide and anilide herbicides range in water solubility from low to moderate (Ksp = 15 to 613 mg/l) (Tables 2 and 10). Three of the compounds are very low to low in volatility (VP = 0.03 × 10⁻⁶ to 4.0 × 10⁻⁶ mmHg) and five of them are moderate to high in volatility (VP = 20 × 10⁻⁶ to 225 × 10⁻⁶ mmHg). Because of their moderate to high vapor pressures, soil-applied compounds, such as alachlor, metolachlor, pronamide, and propachlor, should be

Table 10. Properties of Amide and Anilide Herbicides[a]

Common name	Ksp (mg/l)	VP (mmHg × 10⁻⁶)	Koc	T-1/2 (days)
Alachlor	242	21.7	170	20
Diethatyl-ethyl	105	1.5	1400	21
Diphenamide	260	0.03	210	30
Metolachlor	530	20	150	40
Napropamide	73	4.0	400	40
Pronamide	15	85	1000	30
Propachlor	613	225	150	7
Propanil	225	40	200	2

[a] See footnote in Table 3.

shallow-incorporated into the soil to reduce vapor losses, if no rain falls within five days.

Soil retention of all but one of the amide and anilide herbicides is low (Koc = 150 to 1000) (Table 10). Binding of diethatyl-ethyl is moderate (Koc = 1400). Binding of the compounds in soils has been most highly correlated with soil organic matter and expanding-type clay minerals.[65-67] The amides and anilide herbicides are included with the nonionic pesticides in the Ksp vs. Koc relationship in Figure 1 and the R_f vs. Koc relationship in Figure 2.

Longevity of the amide and anilide herbicides ranges from very short to moderate (T-1/2 = 2 to 40 days) (Tables 2 and 10). Degradation is primarily by microbial degradation by fungi.[43,64,68]

CARBAMATE AND CARBANILATE PESTICIDES

The names and properties of 21 carbamate and carbanilate pesticides are listed in Table 11. Thirteen (62%) of the chemicals range in water solubility from very low to low (Ksp = 0.3 to 89 mg/l) and eight (38%) range in water solubility from moderate to very high (Ksp = 113 to 2.8×10^5 mg/l) (Tables 2 and 11). Volatility of 15 (71%) of the chemicals ranges from very low to low (VP = <0.001×10^{-6} to 10×10^{-6} mmHg). Six have moderate to high volatility (VP = 50×10^{-6} to 232×10^{-6} mmHg).

Sixteen (76%) of the carbamate and carbanilate pesticides range in soil retention from very low to low (Koc = 5 to 1000) and five are moderately retentive (Koc = 1500 to 3000) (Tables 2 and 11). Binding of the chemicals in the soil has been most highly correlated with soil organic matter content.[69-73] Carbamate and carbanilate pesticides are included with the nonionic pesticides in the Ksp vs. Koc relationship in Figure 1 and the R_f vs. Koc relationship in Figure 2.

Longevity of 19 (90%) of the carbamate and carbanilate pesticides ranges from very short to short (T-1/2 = 1 to 30 days) (Tables 2 and 11). Two (10%) of the compounds have moderate longevity (T-1/2 = 50 to 60 days). Microbial degradation by fungi and bacteria was reported to be the primary mode of degradation in soils.[74-77]

Table 11. Properties of Carbamate and Carbanilate Pesticides[a]

Common name	Ksp (mg/l)	VP (mmHg × 10⁻⁶)	Koc	T–1/2 (days)
Aldicarb	6000	200	20	30
Aldoxycarb	1×10^4	90	5	20
Barban	11	0.37	800	30
Bendiocarb	40	5	500	14
Carbaryl	113	1.5	300	10
Carbofuran	700	6.0	30	50
Carbosulfan	0.3	0.3	2000	2
Chlorpropham	89	10	400	30
Desmedipham	8	0.003	1500	30
Fenoxycarb	6.0	0.007	1000	1
Mercaptodimethur	24	0.4	300	30
Methomyl	5.8×10^4	50	40	15
Oxamyl	2.8×10^5	232	10	4
Phenmedipham	4.7	<0.001	3000	30
Propham	250	10	200	20
Propoxur	2000	0.02	30	60
Thiobencarb	30	3	900	21
Thiodicarb	35	70	350	7
Thiophanate	3.0	<0.1	2000	10
Thiophanate-methyl	3.5	<0.07	1800	10
Trimethacarb	58	51	400	20

[a] See footnote in Table 3.

CHLORINATED HYDROCARBON PESTICIDES

The seven chlorinated hydrocarbon pesticides have very low water solubility (Ksp < 10 mg/l) (Tables 2 and 12). Three (43%) of the chemicals are very low in volatility (VP = <1 × 10⁻⁶ mmHg) and four (57%) have moderate to high volatility (VP = 12 × 10⁻⁶ to 217 × 10⁻⁶ mmHg).

Soil retention of the chlorinated hydrocarbons ranges from moderate to high (Koc = 100 to 10⁵) (Table 2 and 12) and has been most highly correlated with soil organic matter content.[78-80] Chlorinated hydrocarbon pesticides are included with the nonionic pesticides in the Ksp vs. Koc relationship in Figure 1.

Longevity of the chlorinated hydrocarbons ranges from short to long (T–1/2 = 21 to 180 days) (Tables 2 and 12) and microbial metabolism, chemical reactions, and photodecomposition were reported to be responsible for degradation of the compounds.[64,81-85] Fungi and bacteria were both reported to be involved depending on the pH of the media.

DINITROANILINE HERBICIDES AND GROWTH REGULATORS

All seven of the dinitroaniline herbicides and growth regulators have very low water solubility (Ksp = 0.019 to 0.50 mg/l) and five (71%) of the compounds are moderate to high in volatility (VP = 30 × 10⁻⁶ to 103 × 10⁻⁶ mmHg) (Tables 2 and 13). Two compounds, flumetralin and prodiamine, are very low in volatility (VP = < 1 × 10⁻⁶ mmHg). The three compounds with the highest

Table 12. Properties of Chlorinated Hydrocarbon Pesticides[a]

Common name	Ksp (mg/l)	VP (mmHg × 10⁻⁶)	Koc	T–1/2 (days)
Chlorothalonil	0.6	0.57	2000	30
Endosulfan	0.32	0.17	10^4	50
Lindane	7.3	95	1100	180
Methoxychlor	0.10	<1	10^5	120
PCNB	0.50	100	5000	21
Toxaphene	0.04	12	10^5	70
Tridiphane	1.8	217	5600	28

[a] See footnote in Table 3.

Table 13. Properties of Dinitroaniline Herbicides and Growth Regulators[a]

Common name	Ksp (mg/l)	VP (mmHg × 10⁻⁶)	Koc	T–1/2 (days)
Benefin	0.10	78	10^4	40
Ethalfluralin	0.30	82	10^4	60
Flumetralin	0.019	<1	10^5	20
Isopropalin	0.08	30	10^5	100
Pendimethalin	0.50	30	10^4	90
Prodiamine	0.03	0.24	10^5	30
Trifluralin	0.3	103	10^4	60

[a] See footnote in Table 3.

VPs, benefin, ethalfluralin, and trifluralin, must be incorporated into the soil to reduce vapor loss and obtain maximum weed control effectiveness.[86]

Soil retention of the dinitroaniline herbicides and growth regulators is high (Koc = 10^4 to 10^5) (Tables 2 and 13), and soil organic matter was the soil constituent most highly correlated with binding in the soil.[87-90] Values of Ksp and Koc for the dinitroanilines lie in the lower right hand quadrant of Figure 1 and are included in the relationship for the nonionic pesticides.

Longevity of the dinitroaniline herbicides and growth regulators ranges from short to moderate (T–1/2 = 20 to 100 days) (Tables 2 and 13). The chemicals are degraded primarily by soil microorganisms, and fungi were the major organisms involved.[86,91-93]

FUMIGANT PESTICIDES

The five fumigant pesticides listed in Table 14 have high to very high water solubility (Ksp = 1620 to 1.3×10^4 mg/l) and would be very mobile in soil if they persisted long enough to be carried downward with percolating water. The compounds, however, are extremely volatile (VP = 20 to 1702 mmHg). To be effective, the chemicals are normally injected into the soil and must diffuse through the soil to control the target pests.

Soil retention of the fumigant pesticides is very low (Koc = 10 to 60) (Tables 2 and 14) and because of the high volatility of the chemicals, plastic

Table 14. Properties of Fumigant Pesticides[a]

Common name	Ksp (mg/l)	VP (mmHg)	Koc	T–1/2 (days)
Chloropicrin	1620	24	60	1
DD	2000	34.5	20	1
1,3-Dichloropropene	2000	27.7	20	2
Methyl bromide	1.3×10^4	1702	22	20
Methylisothiocyanate	7600	20	10	10

[a] See footnote in Table 3.

sheeting is necessary to prevent the chemicals from escaping from the soil too rapidly.[14,94,95] Values for the chemicals lie in the upper left hand quadrant of Figure 1 and the Ksp vs. Koc relationship follows that for nonionic pesticides.

Longevity of fumigant pesticides in soils ranges from very short to short (T–1/2 = 1 to 20 days) mainly because of the high volatility of the chemicals (Table 14). Degradation of the compounds is principally by photodecomposition of the vapors as they escape the soil.[48]

ORGANOPHOSPHATE INSECTICIDES

Of the 34 organophosphate insecticides listed in Table 15 19 (56%) range in water solubility from very low to low (Ksp = 0.001 to 60 mg/l). Fifteen (44%) of the chemicals have moderate to very high (Ksp = 145 to 10^6 mg/l) solubility (Tables 2 and 15). Ten (29%) of the lower solubility organophosphorus insecticides have moderate to high (Koc = 1500 to 10^5) soil retention. Twenty-four (71%) of the higher solubility compounds have very low to low (Koc = 1 to 10^3) soil retention. The Ksp vs. Koc relationship for the organophosphorus insecticides is analogous to that for the nonionic pesticides (Figure 1). R_f values for eight of the organophosphate insecticides lie considerably below the line for the R_f vs. Koc relationship shown in Figure 2. The compounds tend to have lower mobility than soil retention would suggest. This may be because Koc values are for parent compounds only, whereas R_f values may be for the parent and the less mobile metabolites. More on this later.

The organophosphate insecticides vary from very low to very high in volatility (VP = 0.1×10^{-6} mm to 1960×10^{-6} mmHg) (Tables 2 and 15). One third of the compounds are very low to low in volatility (VP = 0.1×10^{-6} to 9×10^{-6} mmHg), one third are moderately volatile (VP = 16×10^{-6} to 92×10^{-6} mmHg), and one third are high to very high (140×10^{-6} to 1960×10^{-6} mmHg) in volatility.

Longevity of the organophosphate insecticides ranges from very short to moderate (T–1/2 = 1 to 120 days (Tables 2 and 15). Eleven (32%) of the organophosphates have very short half-lives and twenty (59%) have short half-lives. Only three (9%) have moderately long half-lives. Organophosphate insecticides are degraded primarily by microbial processes[84,96,97] and bacteria,[98] fungi,[99] and algae[100] have been identified as being involved. Breakdown prod-

Table 15. Properties of Organophosphate Insecticides[a]

Common name	Ksp (mg/l)	VP (mmHg × 10^{-6})	Koc	T–1/2 (days)
Acephate	8.2×10^5	1.7	2	2
Azinphosmethyl	29	16	1000	30
Chlorfenvinphos	145	7.5	1500	30
Chlorpyrifos	0.4	18.7	6000	30
Diazinon	60	60	1000	40
Dicrotofos	1×10^6	140	75	28
Dimethoate	3.2×10^4	8.25	20	7
Disulfoton	25	370	600	30
Ethion	1.1	1.5	10^4	90
Ethoprop	750	349	70	25
Fenamiphos	400	0.45	100	50
Fensulfothion	1540	0.1	50	14
Fenthion	55	50	1000	34
Fonofos	16.9	210	1000	40
Isofenphos	24	9.0	600	90
Malathion	145	20	1800	2
Methamidophos	1×10^6	800	5	6
Methidathion	240	2.5	400	7
Methylparathion	60	20	5000	5
Mevinphos	6×10^5	270	40	3
Monocrotophos	1×10^6	160	1	30
Naled	2000	1960	200	1
Oxydemeton-methyl	1×10^6	60	10	10
Parathion	24	92	5000	14
Phorate	50	1900	1000	7
Phosalone	3.0	0.5	2000	30
Phosmet	25	0.45	700	7
Phosphamidon	1×10^6	55	5	20
Pirimiphos-methyl	5	50	1000	10
Profenophos	28	19	800	4
Sulprofos	0.35	0.63	10^4	120
Temephos	0.001	<10	10^5	30
Terbufos	4	260	1000	14
Tribufos	2.3	1.6	5000	30

[a] See footnote in Table 3.

ucts include P acid metabolites which readily bind to soils similar to that of the organo P acid pesticides (Table 8). This is probably the reason that mobility (R_f) of the chemicals tends to be less than that indicated by Koc values (Figure 2).

PHENYLUREA PESTICIDES

Six phenylurea pesticides are listed in Table 16. The chemicals range from very low to moderate (Ksp = 0.08 to 110 mg/l) in water solubility and low to high in soil retention (Koc = 100 to 10^4). The relationship of Koc vs. Ksp follows that for the nonionic pesticides in Figure 1 and organic matter was the soil constituent most highly correlated with phenylurea binding in soil with clay minerals also contributing.[101-104]

Table 16. Properties of Phenylurea Pesticides[a]

Common name	Ksp (mg/l)	VP (mmHg × 10⁻⁶)	Koc	T–1/2 (days)
Chloroxuron	3.7	0.002	3000	60
Diflubenzuron	0.08	0.001	10^4	10
Diruon	42	0.006	500	90
Fluometuron	110	0.937	100	60
Linuron	81	17	400	60
Siduron	18	0.004	700	90

a See footnote in Table 3.

Volatility of all of the phenylurea pesticides, except linuron, is very low (VP = 1 × 10^{-9} to 0.9 × 10^{-6} mmHg) (Tables 2 and 16); linuron has moderate volatility (VP = 17 × 10^{-6} mmHg).

The phenylurea pesticides have short to moderate longevity (T–1/2 = 10 to 90 days) (Tables 2 and 16). The chemicals are degraded in soil by microbial degradation, and bacteria are the organisms primarily involved.[101,105,106]

PYRETHROID PESTICIDES

The eleven pyrethroid pesticides listed in Table 17 have very low water solubility (Ksp = 0.001 to 0.33 mg/l) and high to very high soil retention (Koc = 10^4 to 10^6). They all lie in the lower right hand quadrant of Figure 1 and along the line for the Ksp vs. Koc relationship for nonionic pesticides.

Volatility of the pyrethroids is very low (VP = 1.5 × 10^{-9} to 0.34 × 10^{-6} mmHg) for all except fenpropathrin which is moderately volatile (VP = 1.1 × 10^{-5} mmHg) (Tables 2 and 17).

Longevity of the pyrethroid pesticides ranges from very short to short (T–1/2 = 3 to 30 days) (Tables 2 and 17). The compounds are relatively unstable in soils and are broken down chemically and biologically.[13]

Table 17. Properties of Pyrethroid Pesticides[a]

Common name	Ksp (mg/l)	VP (mmHg × 10⁻⁶)	Koc	T–1/2 (days)
Bifenthrin	0.1	0.18	10^5	30
Cyfluthrin	0.002	0.016	10^5	30
Cypermethrin	0.004	0.007	10^5	30
Esfenvalerate	0.002	0.01	10^5	30
Fenpropathrin	0.33	11	10^4	3
Fenvalerate	0.002	0.277	10^5	30
Flucythrinate	0.06	0.009	10^5	21
Fluvalinate	0.005	0.1	10^6	30
Lambda-cyhalothrin	0.005	0.0015	10^5	30
Permethrin	0.20	0.34	10^5	20
Tralomethrin	0.001	<0.001	10^5	13

a See footnote in Table 3.

Table 18. Properties of Thiocarbamate Herbicides[a]

Common name	Ksp (mg/l)	VP (mmHg × 10⁻⁶)	Koc	T–1/2 (days)
Butylate	45	13,000	400	14
Cycloate	95	6,225	400	30
Diallate	14	150	800	30
EPTC	375	35,250	200	7
Molinate	970	5,595	190	21
Pebulate	100	35,000	400	14
Triallate	4	120	3,000	50
Vernolate	108	10,425	260	7

[a] See footnote in Table 3.

THIOCARBAMATE HERBICIDES

The most important characteristic of the eight thiocarbamate herbicides listed in Table 18 is the high to very high (VP = 120×10^{-6} to 32520×10^{-6} mmHg) volatility of the compounds. The chemicals must be incorporated into the soil to reduce vapor losses and to put them in intimate contact with weed seeds.[17]

Water solubility of the thiocarbamates ranges from very low to moderate (Ksp = 4 to 970 mg/l) and soil retention ranges from low to moderate (Koc = 190 to 3000) (Tables 2 and 18). Soil organic matter is the soil constituent most highly correlated with retention of the chemicals in soils.[107-109] The relationship of Ksp vs. Koc for the compounds falls on the line for nonionic pesticides in Figure 1.

Longevity for all thiocarbamate herbicides except triallate ranges from very short to short (T–1/2 = 7 to 30 days) (Tables 2 and 18). Triallate has moderate longevity (T–1/2 = 50 days) in soil. The thiocarbamate herbicides are degraded primarily by microorganisms, and bacteria appear to be the major organism involved.[64,110-112]

MISCELLANEOUS, VERY LOW TO LOW WATER SOLUBLE PESTICIDES

Table 19 lists 10 miscellaneous, very low water soluble pesticides (Ksp = 0.006 to 2.6 mg/l) that have moderate to high soil retention (Koc = 10^3 to 10^5). The compounds fall in the lower right hand quadrant of the Ksp vs. Koc relationship of Figure 1 and follow the relationship for nonionic pesticides. They are very low to low in volatility (VP = 0.03×10^{-6} to 3.0×10^{-6} mmHg) and very short to moderate (T–1/2 = 2 to 60 days) in longevity.

MISCELLANEOUS MODERATE TO HIGH WATER SOLUBLE PESTICIDES

Table 20 lists 11 miscellaneous pesticides that have moderate to high (Ksp = 110 to 8400 mg/l) water solubility and very low to low soil retention (Koc

Table 19. Properties of Miscellaneous, Very Low Water
Soluble Pesticides[a]

Common name	Ksp (mg/l)	VP (mmHg × 10⁻⁶)	Koc	T–1/2 (days)
Amitraz	1.0	0.9	1000	2
Chlorothalonil	0.6	0.57	2000	30
Hexythiazox	0.5	0.03	6200	30
Hydramethylnon	0.006	0.02	10^5	10
Methazole	1.5	1	3000	14
Oxadiazon	1.0	<1	3200	60
Oxyfluorofen	0.1	0.19	10^5	35
Oxythioquinox	1.0	0.20	2300	30
Propargite	0.5	3.0	8000	30
Vinclozolin	2.6	2.6	10^4	20

[a] See footnote in Table 3.

Table 20. Properties of Miscellaneous, Moderate to High
Water Soluble Pesticides[a]

Common name	Ksp (mg/l)	VP (mmHg × 10⁻⁶)	Koc	T–1/2 (days)
Ancymidol	650	0.2	120	120
Carboxin	170	0.18	260	35
Clomazone	1100	144	300	24
Cymoxanil	1000	0.6	300	7
Dazomet	3000	6.0	50	7
Dimethipin	3000	1.14	10	10
Ethofumesate	110	0.64	340	15
Metalaxyl	8400	5.6	50	35
Metaldehyde	230	<0.1	240	10
Norea	150	<1	300	20
Pyrazon	400	0.012	200	30

[a] See footnote in Table 3.

= 10 to 340). The compounds lie in the upper left hand quadrant of the Ksp vs. Koc relationship and fall along the line for nonionic pesticides. With the exception of clomazone, they are all very low to low (VP = 0.012×10^{-6} to 6.0×10^{-6} mmHg) in volatility. Clomazone has high volatility (VP = 144×10^{-6}). With the exception of ancymidol, the compounds all have very short to moderate (T–1/2 = 7 to 35 days) longevity. Ancymidol has a long half-life (T–1/2 = 120 days).

SUMMARY

Quaternary N pesticides have very high reactivity with soil colloids and short to very long persistence in soils (Table 21).

Basic pesticides have low to high soil retention in soils and very short to moderate longevity, depending on soil pH (Table 21). The chemicals are bound in greater amounts at lower pH levels than at neutral or higher pH levels.

Table 21. Summary Table of the Relative Behavior of Pesticides in Soil

Category	Parameter[a] Soil reactivity	Longevity
Quaternary N pesticides	vh	s–vl
Basic pesticides	l–h	vs–m
Carboxylic acid herbicides and growth regulators	vl–l	vs–s
Hydroxy acid pesticides	vl–l	vs–s
Aminosulfonyl acid herbicides	vl–l	vs–m
Organic As and P acid pesticides	h–vh	vs–s
Organometallic fungicides	m–vh	s–m
Amide and anilide herbicides	l	vs–m
Carbamate and carbanilate pesticides	vl–l	vs–s
Chlorinated hydrocarbon pesticides	m–h	s–l
Dinitroaniline herbicides and growth regulators	h	s–m
Fumigant pesticides	vl	vs–s
Organophosphate insecticides	vl–l	vs–s
Organophosphate insecticide metabolites	h–vh	vs–s
Phenylurea pesticides	l–h	s–m
Pyrethroid pesticides	h–vh	vs–s
Thiocarbamate herbicides	l–m	vs–s
Misc., very low solubility pesticides	m–h	vs–m
Misc., moderate to high solubility pesticides	vl–l	vs–m

[a] h = high, l = low (soil reactivity) or long (longevity), m = moderate, s = short, v = very.

Carboxylic, hydroxy, and aminosulfonyl acid pesticides have very low to low soil retention and very short to moderate longevity in soils (Table 21). Soil retention and longevity are pH dependent.

Organic As and P acid pesticides, metabolites of organophosphate insecticides, and pyrethroid pesticides have high to very high soil retention and very short to short longevity in soils (Table 21).

Organometallic fungicides have moderate to very high retention in soils and short to moderate longevity (Table 21).

Amide and anilide herbicides have low retention by soils and very short to moderate longevity (Table 21).

Carbamate and carbanilate pesticides and parent organophosphorus insecticides have very low to low soil retention and very short to moderate longevity (Table 21).

Chlorinated hydrocarbon pesticides have moderate to high soil retention and short to long longevity in soils (Table 21).

Dinitroaniline herbicides and growth regulators have high soil retention and short to moderate longevity (Table 21).

Fumigant pesticides have very low soil retention and very short to short longevity (Table 21).

Phenylurea pesticides have low to high soil retention and short to moderate longevity (Table 21).

Thiocarbamate herbicides have low to moderate soil retention and very short to short longevity (Table 21).

Miscellaneous, very low water soluble pesticides have moderate to high soil retention and very short to moderate longevity (Table 21).

Miscellaneous, moderate to high water soluble pesticides have very low to low soil retention and very short to moderate longevity (Table 21).

The inverse relationship between Ksp and Koc was in fair agreement for all but the quaternary N pesticides, basic pesticides with $pK_A > 2$, carboxylic acid pesticides and growth regulators, and organic As and P acid pesticides. No relationship existed for quaternary N pesticides, because of their cationic nature, or the organic As and P acid pesticides because of their tendency to complex to soil clays or to form insoluble precipitates. A special Ksp vs. Koc relationship existed for basic pesticides with $pK_A > 2$, due to the formation and ionic binding of cationic species to organic and mineral soil surfaces at low pH. A special relationship also existed for carboxylic acid herbicides, due to their anionic nature at neutral and high pH. The Koc vs. Ksp relationship was best for lipophilic compounds that were bound almost entirely by organic soil colloids and poorest for compounds that were bound to both organic and mineral surfaces.

ACKNOWLEDGMENTS

The author acknowledges the Water Resources Research Institute and the U.S. Geological Survey (Grant No. 89-0496) for supporting research on pesticides in soil and water.

LITERATURE CITED

1. Sawhney, B. L. and Brown, K.; Eds. *Reactions and Movement of Organic Chemicals in Soils;* Soil Science Society of America Special Publication No. 22; Soil Science Society of America, Inc., Madison, WI, 1989.
2. Weber, J. B. *Appl. Plant Sci.* 1991, 5, 28–41.
3. Weber, J. B. and Miller, C. T. In *Reactions and Movement of Organic Chemicals in Soils;* Sawhney, B. L. and Brown, K.; Eds.; Soil Science Society Special Publication No. 22, Soil Science Society of America, Inc., Madison, WI, 1989, pp. 305–334.
4. Breth, S. A. and Stelly, M.; Eds. *Pesticides and Their Effects on Soil and Water,* Soil Science Society of America, Inc., Madison, WI, 1966.
5. Guenzi, W. D.; Ed. *Pesticides in Soil and Water;* Soil Science Society of America, Inc., Madison, WI, 1974.
6. Hartley, G. S. and Graham-Bryce, I. J. *Physical Principles of Pesticide Behavior;* Academic Press, Inc., New York, NY, 1980, Vol. 1.
7. *Agrochemicals Handbook,* 2nd ed., The Royal Society of Chemistry, Cambridge, England, 1990.
8. Camper, N. D.; Ed. *Research Methods in Weed Science,* Southern Weed Science Society, Inc., Champaign, IL, 1986.

9. *Farm Chemicals Handbook.* Meister Publishing Co., Willoughby, OH, 1991.
10. Frear, D. E. H.; Ed. *Pesticide Index;* 4th edition; College Science Publishers, State College, PA, 1969.
11. *Herbicide Handbook.* Weed Science Society of America, Inc., Champaign, IL, 1989.
12. Verschueren, K. *Handbook of Environmental Data on Organic Chemicals;* Von Nostrand Reinhold Co., New York, NY, 1983.
13. Buchel, K. H.; Ed. *Chemistry of Pesticides;* Holmwood, G.; Translator; John Wiley and Sons, Inc., New York, NY, 1983.
14. Cremlyn, R. *Pesticides: Preparation and Mode of Action,* John Wiley and Sons, Inc., New York, NY, 1978.
15. Melnikov, N. N. *Chemistry of Pesticides;* Gunther, F. A. and Gunther, J. D.; Eds.; Busbey, R. L.; Translator; Springer-Verlag, Inc., New York, NY, 1971.
16. Worthing, C. R.; Ed. *The Pesticide Manual;* 6th edition; The British Crop Protection Council, London, England, 1979.
17. Weber, J. B. In *Fate of Organic Pesticides in the Aquatic Environment;* Gould, R. F.; Ed.; Advances in Chemistry Series No. 111; American Chemical Society, Washington, DC, 1972, pp. 55–120.
18. Weber, J. B. *Weed Sci.* 1980, 28, 478–483.
19. Weber, J. B. and Weed, S. B. In *Pesticides in Soil and Water;* Guenzi, W. D.; Ed.; Soil Science Society of America, Inc., Madison, WI, 1974, pp. 223–256.
20. Weed, S. B. and Weber, J. B. In *Pesticides in Soil and Water;* Guenzi, W. D.; Ed.; Soil Science Society of America, Inc., Madison, WI, 1974, pp. 39–66.
21. Wauchope, R. D. *Interim Pesticide Properties Data Base;* Version 1.0; U.S. Department of Agriculture, Agricultural Research Service, South East Watershed Research Lab., Tifton, GA, 1988.
22. Wauchope, R. D.; Butler, T. M.; Hornsby, A. G.; Augustijn-Beckers, P. W.; and Burt, J. P.; *CES Pesticide Properties Database;* Soil Conservation Service/Agricultural Research Service; U.S. Department of Agriculture, South East Watershed Research Lab., Tifton, GA, 1991.
23. Best, J. A.; Weber, J. B.; and Weed, S. B. *Soil Sci.* 1972, 114, 444–450.
24. Calderbank, A. and Slade, P. In *Herbicides: Chemistry, Degradation, and Mode of Action,* Kearney, P. C. and Kaufman, D. D.; Eds.; Marcel Dekker Inc., New York, NY, 1976, Vol. 2, pp. 501–540.
25. Summers, L. A. *The Bipyridylium Herbicides;* Academic Press, Inc., New York, NY, 1980.
26. Weber, J. B. and Coble, H. D. *J. Agr. Food Chem.* 1968, 16, 475–478.
27. Weber, J. B. *Spectrochimica Acta* 1967, 23A, 458–461.
28. Gunther, F. A.; Ed. *The Triazine Herbicides;* Springer-Verlag, Inc., New York, NY, 1970.
29. Weber, J. B. *Res. Rev.* 1970, 32, 93–130.
30. Weber, J. B. *Soil Sci. Soc. Am. Proc.* 1970, 34, 401–404.
31. Weber, J. B. *J. Agr. Food Chem.* 1982, 30, 584–588.
32. Weber, J. B.; Shea, P. J.; and Weed, S. B. *Soil Sci. Soc. Am. Proc.* 1986, 50, 582–588.
33. Weber, J. B.; Weed, S. B.; and Ward, T. M. *Weed Sci.* 1969, 17, 417–421.
34. Woolson, E. A. In *Herbicides: Chemistry, Degradation, and Mode of Action;* Kearney, P. C. and Kaufman, D. D.; Eds.; Marcel Dekker, Inc., New York, NY, 1976, Vol. 2, pp. 741–776.

35. Armstrong, D. E. and Chesters, G. *Soil Sci. Soc. Am. Proc.* 1967, 31, 61–66.
36. Best, J. A. and Weber, J. B. *Weed Sci.* 1974, 22, 364–373.
37. Skipper, H. D.; Gilmour, C. M.; and Furtick, W. R. *Soil Sci. Soc. Am. Proc.* 1967, 31, 653–656.
38. Hill, I. R. and Wright, S. J. L.; Eds. *Pesticide Microbiology,* Academic Press, Inc., New York, NY, 1978.
39. Frear, D. S. In *Herbicides: Chemistry, Degradation, and Mode of Action;* Kearney, P. C. and Kaufman, D. D.; Eds.; Marcel Dekker, Inc., New York, NY, 1976, Vol. 2, pp. 541–607.
40. Hamaker, J. W. and Thompson, J. M. In *Organic Chemicals in the Soil Environment;* Goring, C. A. I. and Hamaker, J. W.; Eds.; Marcel Dekker, Inc., New York, NY, 1972, Vol. 1, pp. 49–143.
41. Shea, P. J.; Weber, J. B.; and Overcash, M. R. *Res. Rev.* 1983, 87, 1–41.
42. Weber, J. B.; Perry, P. W.; and Upchurch, R. P. *Soil Sci. Soc. Am. Proc.* 1965, 29, 678–688.
43. Cripps, R. E. and Roberts, T. R. In *Pesticide Microbiology;* Hill, I. R. and Wright, S. J. L.; Eds.; Academic Press, New York, NY, 1978, pp. 669–730.
44. Foy, C. L. In *Herbicides: Chemistry, Degradation, and Mode of Action;* Kearney, P. C. and Kaufman, D. D.; Eds.; Marcel Dekker, Inc., New York, NY, 1975, Vol. 1, pp. 399–452.
45. Foy, C. L. In *Herbicides: Chemistry, Degradation, and Mode of Action;* Kearney, P. C. and Kaufman, D. D.; Eds.; Marcel Dekker, Inc.; New York, NY, 1976, Vol. 2, pp. 777–813.
46. Loos, M. A. In *Herbicides: Chemistry, Degradation, and Mode of Action;* Kearney, P. C. and Kaufman, D. D.; Eds.; Marcel Dekker, Inc., New York, NY, 1975, Vol. 1, pp. 1–28.
47. Tweedy, B. G. and Houseworth, L. D. In *Herbicides: Chemistry, Degradation and Mode of Action;* Kearney, P. C. and Kaufman, D. D.; Eds.; Marcel Dekker, Inc., New York, NY, 1976, Vol. 2, pp. 815–833.
48. Crosby, D. G. In *Herbicides: Chemistry, Degradation, and Mode of Action;* Kearney, P. C. and Kaufman, D. D.; Eds.; Marcel Dekker, Inc., New York, NY, 1976, Vol. 2, pp. 835–890.
49. Gardiner, J. A. In *Herbicides: Chemistry, Degradation, and Mode of Action;* Kearney, P. C. and Kaufman, D. D.; Eds.; Marcel Dekker, Inc., New York, NY, 1975, Vol. 1, pp. 293–321.
50. Kaufman, D. D. In *Herbicide: Chemistry, Degradation, and Mode of Action;* Kearney, P. C. and Kaufman, D. D.; Eds.; Marcel Dekker, Inc., New York, NY, 1976, Vol. 2, pp. 665–707.
51. Beyer, E. M., Jr.; Duffy, M. J.; Hay, J. V.; and Schlueter, D. D. In *Herbicides: Chemistry, Degradation, and Mode of Action;* Kearney, P. C. and Kaufman, D. D.; Eds.; Marcel Dekker, Inc., New York, NY, 1988, Vol. 3, pp. 117–189.
52. Liu, S. L. and Weber, J. B. *Proc. South. Weed Sci. Soc.* 1985, 38, 465–474.
53. Mersie, W. and Foy, C. L. *J. Agr. Food Chem.* 1986, 34, 89–93.
54. Dickens, R. and Hiltbold, A. E. *Weeds* 1967, 15, 299–304.
55. Duke, S. D. In *Herbicides: Chemistry, Degradation, and Mode of Action,* Kearney, P. C. and Kaufman, D. D.; Eds.; Marcel Dekker Inc., New York, NY, 1988, Vol. 3, pp. 1–70.
56. Grossbard, E. and Atkinson, D.; Eds. *The Herbicide Glyphosate;* Butterworths, London, England, 1985.

57. Hiltbold, A. E.; Hajek, B. F.; and Buchanan, G. A. *Weed Sci.* 1974, 22, 272–275.
58. Woolson, E. A.; Ed. *Arsenical Pesticides;* American Chemical Symposium Series 7; American Chemical Society, Washington, DC, 1975.
59. Tortenson, L. In *The Herbicide Glyphosate;* Grossbard, E. and Atkinson, D.; Eds.; Butterworths, London, England, 1985, pp. 137–150.
60. Hiltbold, A. E. In *Arsenical Pesticides;* Woolson, G. A.; Ed.; Advance in Chemistry Symposium Series No. 7; American Chemical Society, Washington, DC, 1975, pp. 53–69.
61. Von Endt, D. W.; Kearney, P. C.; and Kaufman, D. D. *J. Agr. Food Chem.* 1968, 16, 17–20.
62. Van der Kerk, G. J. M. In *Pesticide Chemistry in the 20th Century;* Plimmer, J. R.; Ed.; American Chemical Symposium Series No. 37, American Chemical Society, Washington, DC, 1977, pp. 123–152.
63. Dubey, H. D. and Rodriguez, R. L. *J. Agr. Univ. Puerto Rico* 1974, 34, 78–86.
64. Kaufman, D. D. In *Pesticides in Soil and Water;* Guenzi, W. D.; Ed.; Soil Science Society of America, Inc., Madison, WI, 1974, pp. 133–202.
65. Kozak, J.; Weber, J. B.; and Sheets, T. J. *Soil Sci.* 1983, 136, 94–101.
66. Parochetti, J. V. *Weed Sci.* 1973, 21, 157–159.
67. Peter, C. J. and Weber, J. B. *Weed Sci.* 1985, 33, 874–881.
68. Jaworski, E. G. In *Herbicides: Chemistry, Degradation, and Mode of Action;* Kearney, P. C. and Kaufman, D. D.; Eds.; Marcel Dekker, Inc., New York, NY, 1985, Vol. 1, pp. 349–376.
69. Harris, C. I. and Sheets, T. J. *Weeds* 1965, 13, 215–219.
70. Leenheer, J. A. and Alrichs, J. L. *Soil Sci. Soc. Am. Proc.* 1971, 35, 700–705.
71. Roberts, H. A. and Wilson, B. J. *Weed Res.* 1965, 5, 348–350.
72. Scott, D. C. and Weber, J. B. *Soil Sci.* 1967, 104, 151–158.
73. Upchurch, R. P. and Mason, D. D. *Weeds* 1962, 10, 9–14.
74. Bartha, R. and Pramer, D. *Bull. Environ. Contam. Toxicol.* 1969, 4, 240–245.
75. Kaufman, D. D. and Kearney, P. C. *Appl. Microbiol.* 1965, 13, 443–446.
76. Still, G. G. and Herrett, R. A. In *Herbicides: Chemistry, Degradation, and Mode of Action;* Kearney, P. C. and Kaufman, D. D.; Eds.; Marcel Dekker, Inc., New York, NY, 1976, Vol. 2, pp. 609–664.
77. Wright, S. J. L. and Forey, A. *Soil Biol. Biochem.* 1972, 4, 207–213.
78. Adams, R. S., Jr. and Li, P. *Soil Sci. Soc. Am. Proc.* 1971, 35, 78–81.
79. Edwards, C. A.; Beck, S. D.; and Lichtenstein, E. P. *J. Econ. Entomol.* 1957, 50, 622–626.
80. Kay, B. D. and Elrick, D. E. *Soil Sci.* 1967, 104, 314–322.
81. Hill, D. W. and McCarty, P. L. *J. Water Pollut. Contr. Fed.* 1967, 39, 1259–1277.
82. Lichtenstein, E. P. and Schultz, K. R. *J. Econ. Entomol.* 1960, 53, 192–197.
83. Martens, R. *Schriftenr. Ver. Wasser-Boden-Lufthyg. Berlin-Dahlem* 1972, 37, 167–173.
84. Matsumura, F. and Benezet, H. J. In *Pesticide Microbiology;* Hill, I. R. and Wright, S. J. L.; Eds.; Academic Press, Inc., New York, NY, 1978, pp. 623–667.
85. Mendel, J. L.; Klein, A. K.; Chen, J. T.; and Walton, M. S. *J. Assoc. Off. Agr. Chem.* 1967, 50, 897–903.
86. Weber, J. B. *Weed Technol.* 1990, 4, 394–406.
87. Carringer, R. D.; Weber, J. B.; and Monaco, T. J. *J. Agr. Food Chem.* 1975, 23, 568–572.

88. Harrison, G. W.; Weber, J. B.; Baird, J. V. *Weed Sci.* 1976, 24, 120–126.

89. Jacques, G. L. and Harvey, R. G. *Weed Sci.* 1979, 27, 450–455.

90. Peter, C. J. and Weber, J. B. *Weed Sci.* 1985, 33, 861–867.

91. Golab, T. and Althaus, W. A. *Weed Sci.* 1975, 23, 165–171.

92. Hamdi, Y. A. and Tewfik, M. S. *Acta Microbial. Pol. Ser. B.* 1969, 1, 83–84.

93. Laanio, T. L.; Kearney, P. C.; and Kaufman, D. D. *Pestic. Biochem. Physiol.* 1973, 3, 271–277.

94. Call, F. *J. Sci. Food Agr.* 1957, 8, 137–142.

95. Letey, J. and Farmer, W. J. In *Pesticides in Soil and Water;* Guenzi, W. D.; Ed.; Soil Science Society of America, Inc., Madison, WI, 1974, pp. 67–97.

96. Ahmed, M. K. and Casida, J. E. *J. Econ. Ent.* 1958, 51, 59–63.

97. Konrad, J. G.; Chesters, G.; and Armstrong, D. E. *Soil Sci. Soc. Am. Proc.* 1969, 33, 259–262.

98. Yasuno, M.; Hirakoso, S.; Sasa, M.; and Uchida, M. *Jap. J. Exp. Med.* 1965, 35, 545–563.

99. Sethunathan, N. and MacRae, I. C. *J. Agr. Food Chem.* 1969, 17; 221–225.

100. Sato, R. and Kubo, H. *Adv. Water Pollut. Res.* 1964, 1, 95–99.

101. Geissbuhler, H.; Martin, H.; and Voss, G. In *Herbicides: Chemistry, Degradation, and Mode of Action;* Kearney, P. C. and Kaufman, D. D., Eds.; Marcel Dekker, Inc., New York, NY, 1975, Vol. 1, pp. 209–291.

102. Graves, R. *Can. J. Soil Sci.* 1975, 55, 127–135.

103. Hance, R. J. *Weed Res.* 1969, 9, 108–113.

104. Kozak, J. and Weber, J. B. *Weed Sci.* 1983, 31, 368–372.

105. Engelhardt, G.; Wallnofer, P. R.; and Plapp, R. *Appl. Microbiol.* 1971, 22, 284–288.

106. Wallnofer, P. *Weed Res.* 1969, 9, 333–339.

107. Ashton, F. M. and Sheets, T. J. *Weeds* 1959, 7, 88–91.

108. Deming, J. M. *Weeds* 1963, 11, 91–93.

109. Koren, E.; Foy, C. L.; and Ashton, F. M. *Weed Sci.* 1968, 16, 172–174.

110. Fang, S. C. In *Herbicides: Chemistry, Degradation, and Mode of Action;* Kearney, P. C. and Kaufman, D. D.; Eds.; Marcel Dekker, Inc., New York, NY, 1975, Vol. 1, pp. 323–348.

111. Kaufman, D. D. *J. Agr. Food Chem.* 1967, 15, 582–591.

112. MacRae, I. C. and Alexander, M. *J. Agr. Food Chem.* 1965, 13, 72–76.

CHAPTER 3

Mobility of Pesticides in Field Lysimeters

J. B. Weber and K. E. Keller

Soil column field lysimeters of undisturbed soils are as close to natural field conditions as we can safely get using ^{14}C-labeled pesticides. Tritium mobility studies showed that water moved horizontally down the soil cores. Mass balance studies of the distribution and dissipation of three herbicides in a Coastal Plains loamy sand (Dothan series, *Plinthic Kandiudult*) soil showed that ^{14}C-metolachlor was more mobile than ^{14}C-primisulfuron and ^{14}C-atrazine, respectively. Liming the soil increased the mobility and longevity of atrazine. Growing plants (soybean) or bermuda sod in the lysimeters reduced and wheat straw mulch increased ^{14}C-metolachlor mobility as compared with fallow lysimeters. Herbicide movement in 15- and 20-cm-diameter columns was similar, but movement through 10-cm-diameter columns was slower. Leachate collectors attached to the bottoms of lysimeters reduced herbicide movement slightly as compared with lysimeters with no leachate collectors.

The movement of water and nutrients through soils using lysimeters has been studied for nearly three centuries.[1] Joffe[2] described lysimeters in the form of large concrete or metal boxes filled with soil and equipped with leachate collection devices to measure percolating water and nutrients. Musgrave[3] described the use of 1-m-diameter steel cylinders which were slipped over 1-m deep undisturbed soil columns and equipped with leachate collection devices at the bottom. Kilmer et al.,[4] Van Bavel,[5] and Tackett et al.[6] drove steel cylinders 66 to 96 cm in diameter by 112 to 275 cm in length into the soil to obtain undisturbed soil cores for use in studying transpiration, evapotranspiration, and moisture flow in a controlled environment. More recently, investigators have used column lysimeters to study the movement and dissipation of pesticides,[7-12] gases,[13] tritiated water,[14,15] nitrates,[2,16] and halides,[14,17,18] in both disturbed and undisturbed soils.

43

Ideally, lysimeters should simulate natural field conditions as much as possible by having the following characteristics:

1. Undisturbed, representative soil profiles large enough and deep enough to provide for normal plant growth,
2. Thin enough border so as not to significantly affect soil density (<20 gauge) when being installed and constructed of material that will readily transmit normal soil heat flux (metallic) and that will bind with soil constituents (will rust or oxidize) to prevent wall flow of water and not be toxic to plants and soil microflora (nongalvanized), if plants are to be grown, and
3. Maintain capillary connections at the bottom so as to have moisture tensions comparable to the rest of the field.

Jordan[19] described a tension-free lysimeter that consisted of a modified Ebermayer trough installed through a horizontal tunnel into the soil profile from a trench, the trough having been filled with a fiberglass screen lined with glass wool and soil, and wedged against the top of the tunnel so as to prevent surface tension from developing at the soil-trough interface. The trough was equipped with tubes to carry percolating water to a covered container located in the trench. Many other investigators have attached leachate collection devices to the bottom of their lysimeters.

Kubiak et al.[20] recently studied the movement of several herbicides through column lysimeters and found herbicide distribution through disturbed and undisturbed soil profiles to be relatively similar.

The objectives of our studies were to utilize field column lysimeters to measure the movement and dissipation of pesticides, nutrients, and water through undisturbed soil cores as influenced by various parameters, including pesticide properties, crop plants, soil pH, soil surface cover, column diameter, and the presence or absence of leachate collectors. The data obtained are being used to validate existing models which predict the mobility of pesticides in soils.

MATERIALS AND METHODS

Lysimeter Installation

Cold-rolled 18-gauge steel columns (20-cm i.d. by 92 cm long) were driven 30 cm apart into a tilled Dothan soil (fine-loamy siliceous, thermic *Plinthic Kandiudult*) by an inverted tractor-mounted postdriver in 1988 and 1989 as described in Patents 4, 166, 720,[21] and 5,009, 112.[22] In 1989, two 10-cm and two 15-cm i.d. columns were also installed to study the effects of column diameter on herbicide movement. Prior to installation of the columns, a metal plate was placed atop each column to prevent damage to the column from the impact of the postdriver. To protect against runoff, 3 cm of the column remained above the soil surface. After installation a 1.2-m wide by 1.5-m deep trench secured with 1.9-cm thick by 1.2×2.4-m sheets of marine plywood and treated 5×10-cm wood planks was established parallel to and 30 cm from the columns. The trench enabled access to holes excavated under selected columns for continuous collection of soil leachate. Foil-lined funnels (20-cm i.d.) and 1-l glass jars were positioned under selected columns to collect all water passing through the columns. The remaining columns maintained soil contact at the 90 cm depth.

Chemical Preparation and Application

On June 30, 1988, a mixture of ^{14}C-metolachlor and formulated metolachlor was applied to bare soil (fallow) and bare soil seeded with soybeans. On June 30, 1989, a mixture of ^{14}C-atrazine and formulated atrazine, ^{14}C-metolachlor, and formulated metolachlor, or ^{14}C-primisulfuron, were applied in 20 ml of water in a cross-hatch pattern to the bare soil surface of selected columns. ^{14}C-metolachlor was also applied to bermuda sod and mulch-covered soil columns. The quantities of ^{14}C-labeled herbicides applied to the columns were as follows: 0.57 MBq atrazine (sp. act. [specific activity] $= 0.72 \times 10^6$ MBq kg^{-1}), 0.56 MBq metolachlor (sp. act. $= 0.71 \times 10^6$ MBq kg^{-1}) and 0.54 MBq primisulfuron (sp. act. $= 2.08 \times 10^6$ MBq kg^{-1}). Radiochemical purity of the chemicals was greater than 97%. Application solutions were equivalent to 1.12 kg ai ha^{-1}, 2.24 kg ai ha^{-1}, and 80 g ai ha^{-1} (4 times recommended rate) for atrazine, metolachlor, and primisulfuron, respectively. The 4 times recommended rate was needed to optimize the accuracy of detection of radiation in 1-g soil samples. In 1989, finely ground agricultural limestone (CaCO$_3$) was applied to two columns at rates of 2240 kg/ha and worked into the soil to a depth of 7.6 cm. ^{14}C-atrazine was applied to these columns. GLP (good laboratory practice) and QA (quality assurance) by a nonrelative were used throughout the study. Duplicate treatments were used in all cases. Control columns (no herbicides applied) were also installed for radiation background and soil property determinations.

Vegetative Establishment

Four soybean (*Glycine max* [L.] Merr. variety 'Coker') seeds were planted in each of eight columns and, once plants were established, thinned later to one plant per column. Bermuda grass sod (*Cynodon dactylon* [L.] Pers.) was established in two columns and wheat straw (*Triticum aestivum* [L.]) was applied at a rate equivalent to 3.2 t ha^{-1} to two columns. Vegetative establishment was made prior to herbicide application.

Leaching and Leachate Collection

Supplemental water was added to each column on a weekly basis such that each column received an amount of water equivalent to the 10-year average rainfall for that same time period. Leachate was monitored daily and collected weekly from 20 columns with and without soybeans. Leachate was quantified and radioassayed for ^{14}C, immediately or placed in freezers at $-20°C$ and analyzed within 1 month. Two 1-ml aliquots of leachate were placed into scintillation vials containing 15 ml of scintillation fluid (Scintiverse E, Fisher Scientific Company, 50 Fadem Road, Springfield, NJ 07081) and ^{14}C determined by radioassay in a liquid scintillation spectrophotometer (Packard TRI-CARB Liquid Scintillation Analyzer, Model 200CA, Packard Instrument Co., 2200 Warrenville Road, Downers Grove, IL 60515). Samples for herbicide/ metabolite analysis were stored in glass containers at $-20°C$, and analyzed within 2 months.

Lysimeter Removal and Sectioning

Zero day soil samples of the 0 to 7.5 cm depth were taken and analyzed in triplicate (for each treatment) and used as an index of the amount applied and the amount recovered, which was 100% in all cases. Duplicate columns were removed at 1, 2, 3, 4, 6, and 12 months for fallow columns and 1, 2, 3, and 4 months for columns with soybean plants. Duplicate columns with Bermuda grass and wheat straw were removed at 3 months. After removing soil from around the columns, the columns were lifted by a winch to the soil surface. Column ends were secured with plastic and transported vertically to the laboratory. Columns were cut lengthwise and the soil removed in twelve 7.5-cm depth increments, weighed, mixed in a 7-cm deep by 30 by 46-cm-stainless steel pan with stainless steel spatulas placed in polyethylene bags, labeled, and sealed. Sampling equipment was washed with deionized water and rinsed with acetone between each sample. Samples for herbicide/metabolite analysis were stored in polyethylene bags at $-20°C$, and analyzed within 6 months.

Soil and Vegetation Analysis

Water content of soil samples from each section was determined by oven-drying small samples at $105°C$ for 8 h. Four 1-g subsamples from each section were combusted in a biological oxidizer (OX-300 Automated Biological Oxidizer, R. J. Harvey Instrument Co., 123 Patterson St., Hillsdale, NJ 07642). The $^{14}CO_2$ evolved was trapped in carbon scintillation cocktail (OX-161 Carbon 14 Cocktail, R. J. Harvey Instrument Co., 123 Patterson St., Hillsdale, NJ 07642) and subsequently assayed in a liquid scintillation counter. Biological oxidizer efficiency, which was checked weekly, was always greater than 85% and was in most cases greater than 95%. Soil from each 7.5 cm depth of two columns was characterized for humic matter, organic matter, organic carbon, and moisture contents, particle size distribution, pH, undisturbed bulk density, and cation exchange capacity (CEC). Humic matter contents were determined (N.C. State Dep. Agric., Agronomic Div., Blue Ridge Rd., Raleigh, NC 27607) by using the NaOH/DTPA-alcohol extraction method,[23] and organic matter contents were determined (A&L Agricultural Laboratory Inc., 13611 "B" St., Omaha, NE 68114) using the chromic acid colorimetric method.[24] Organic carbon content was determined by the carbon train method.[25] Particle size analyses were performed using the hydrometer method.[26] Soil pH was measured in a 1:1 (w/w) soil:water mixture with a glass electrode pH meter and buffer solutions as references. CEC of the soil was determined by using the Mehlich 3 extracting solution.[27] Undisturbed bulk density was determined using the core method.[28] Vegetative fractions were separated from the soil, washed, dried at 80°C, and ground in a stainless steel micro-Wiley mill. Percent of applied ^{14}C in the vegetative portions were determined by combusting two 0.1-g subsamples and trapping the evolved $^{14}CO_2$ in carbon trapping scintillation cocktail. ^{14}C was determined by radioassay in a liquid scintillation counter. ^{14}C values for plant, soil, and water were corrected by subtracting background values (approximated 35 dpm g^{-1}) from control columns in each case.

For selected samples, ^{14}C-atrazine was extracted by shaking 100 g of air dry soil in 200 ml of methanol for 5 h, filtering under vacuum, and rotoevaporating to a volume of 0.5 ml. ^{14}C-metolachlor was extracted in the same manner except shaking was 1 h. ^{14}C-primisulfuron was extracted similarly by shaking 100 g of moist soil (moisture content determined) in 200 ml of acetonitrile for 1 h. ^{14}C-herbicide extracts were spotted onto methanol-pretreated Whatman LK-5F (linear K) silica gel plates and R_f values for parent and metabolites of atrazine, metolachlor, and primisulfuron were compared using the following, respective, solvent systems: ethyl acetate:volume 1:1; chloroform:methanol:formic acid:water 75:20:4:2; and methylene chloride:toluene 1:1.

Water Flow in the Soil Cores

To evaluate the flow pattern of water through the soil cores, 3.7 MBq of tritium (specific activity = 3.7×10^3 MBq ml^{-1} from New England Nuclear) was applied in 825 ml (2.5 cm) of water to the surface of each of two columns, immediately followed by 1650 ml (5.1 cm) of water on July 1, 1989. The tops of the columns were covered with plastic and allowed to drain for 3 weeks. Approximately 1.5% of the applied tritium was found in leachate from the columns, so the applied tritium and water just displaced the original water held in the soil cores. The columns were removed from the soil, ends capped, and the column stored in a vertical position for 6 weeks. The columns were then cut into three sections at 25, 50, and 75 cm below the soil surface. Soil samples of 1.7 cm^2 were taken across the entire diameter of the soil core, from wall to wall, of each section with a cork borer. Each soil sample was placed in a glass vial, mixed thoroughly, transferred and combusted in a biological oxidizer, ^{3}H was trapped in cocktail, and assayed in a liquid scintillation counter.

Climatic Conditions

Air temperature, soil temperature, and rainfall were recorded by the Central Crops Research Station Weather Station, Clayton, NC, NOAA Site No. 1820. Irrigation and pan evaporation data was recorded weekly at the site.

Parameters Measured and Calculated

^{14}C determinations were made on plant, soil, and leachate samples and all values converted to % of applied herbicide after correction by subtracting background. Leachate volumes were measured weekly and reported in liters (l). Soil moisture was determined at each sampling and reported on a dry-weight (g/g) basis. ^{14}C-herbicide mobility was compared using a computed R_f value. Distribution of each herbicide in the soil profile was expressed as a mean computed R_f value, as calculated by multiplying the fraction of ^{14}C recovered at each depth times the mean soil depth, summing the values for each column, normalizing to 100% recovered, and dividing by the maximum distance that all of the chemical could have moved (88 cm). A sample calculation is shown in Table 1.

RESULTS AND DISCUSSION

Properties of the Herbicides

Names and properties of the three herbicides studied are as follows: Atrazine, 6-chloro-N-ethyl-N'-(1-methylethyl)-1,3,5-triazine-2,4-diamine, is a weakly basic compound (pK$_A$ = 1.7),[29] with low water solubility (33 mg l^{-1} at pH 7 and 20°C),[30] very low vapor pressure (3.0×10^{-7} mmHg at 20°C),[30] and moderate leaching potential (PLP = 0.42).[31]

Metolachlor, 2-chloro-N-(2-ethyl-6-methylphenyl)-N-(2-methoxy-1-methylethyl)acetamide, is a nonionic compound, with moderate water solubility (530 mg l^{-1} at 20°C),[30] moderate vapor pressure (1.3×10^{-5} mmHg at 20°C),[30] and moderate leaching potential (PLP = 0.78).[31]

Primisulfuron, 3-[4,6-bis(difluoromethoxy)-pyrimidine-2-yl]-1-(2-methoxycarbonylphenylsulfonyl)urea, is a weakly acidic compound (pK$_A$ = 5.1), with low water solubility (70 mg l^{-1} at pH 7 and 20°C), and very low vapor pressure ($<7.5 \times 10^{-12}$ mmHg) (Technical Release, CIBA-GEIGY Corp., Greensboro, NC, 1988), and very low leaching potential (PLP = <0.01).[31]

Table 1. Example of R_f Calculation for ^{14}C-Herbicide Distribution in Soil Core

Soil Section (cm)	Mean soil depth (D) (cm)	Fraction of ^{14}C recovered (F)	Depth × Fraction (D × F)
0–7.5	4	0.232	0.928
7.5–15.0	11	0.084	0.924
15.0–22.5	19	0.036	0.684
22.5–30.0	27	0.013	0.351
30.0–37.5	34	0.004	0.136
37.5–45.0	42	0.002	0.084
45.0–52.5	49	0.001	0.049
52.5–60.0	57	0.001	0.057
60.0–67.5	65	0.001	0.065
67.5–75.0	72	0.000	0.000
75.0–82.5	80	0.000	0.000
82.5–90.0	88	0.000	0.000
	Total	0.374	3.278

Normalized to fraction of 1.000 recovered = 3.278 ÷ 0.374 = 8.765
Maximum R_f = (Max.D)(F) ÷ (Max.D)(Max.F) = (88)(1.000) ÷ (88)(1.000) = 1.00
Minimum R_f = (Min.D)(F) ÷ (Max.D)(Max.F) = (4)(1.000) ÷ (88)(1.000) = 0.04
Computed R_f for example = ΣObs'd. D × F ÷ (Max.D)(MaxF.) = 8.765 ÷ (88)(1.000) = 0.10

Properties of the Soils

The Dothan series is a well drained, upland, Coastal Plains soil underlain by a red to variegated red, yellow, and gray weathering zone containing plinthite or soft iron nodules. Properties of the 0 to 90-cm soil profile are given in Table 2. The texture of the profile ranged from loamy sand to sandy clay loam, as kaolinite clay content increased with depth from 6% in the surface to 29% at 90 cm. The organic components of the soil were low and decreased rapidly with depth. CEC of the profile was relatively uniform ranging from 1.9 to 4.2 cmol Kg^{-1} due to the increasing clay content and decreasing organic content with depth. Undisturbed bulk density of the soil profile was relatively uniform at 1.6 g cm^{-1}. Soil pH of the profile ranged from slightly acid (pH 6.2) at the surface to very acid (pH 4.3 to 4.4) in the subsoil. Soil moisture increased with depth from 5.8 to 17.7% as clay content increased.

Climate Conditions

Mean monthly air and soil temperatures for 1988 and 1989 for the site are shown in Figure 1. Temperatures were relatively similar for all years. Soil temperatures for 1988 were substantially cooler than 1989 during the period of February to April due to the cooler air temperatures during January of 1988.

Figure 2A depicts the mean monthly rainfall plus irrigation (1.3 cm of water added at the end of the week if no rainfall occurred during the week for the

Table 2. Properties of Dothan Loamy Sand (Means of Two Replications)

Mean soil depth (cm)	Particle size (%)		Textural class[a]	Organic matter (%)	Humic matter (%)	Organic carbon	C.E.C. (cmol kg⁻¹)	Bulk density (g cm⁻¹)	pH	Soil moisture[b] (%)
	silt	clay								
4	10	6	ls	1.1	0.4	0.4	3.3	1.5	6.2	5.8
11	8	7	ls	0.9	0.4	0.2	2.9	1.7	5.6	6.8
19	9	8	ls	0.8	0.4	0.2	2.3	1.7	5.2	6.3
27	9	9	ls	0.5	0.3	0.1	1.9	1.7	5.1	6.2
34	9	9	ls	0.5	0.1	0.1	1.9	1.7	5.1	7.5
42	9	11	sl	0.5	0.1	0.1	2.4	1.6	5.0	8.4
49	9	15	sl	0.5	0.1	0.1	2.5	1.7	4.8	9.6
57	8	19	sl	0.5	0.1	0.1	3.0	1.6	4.6	11.5
65	9	21	scl	0.5	0.1	0.1	3.6	1.7	4.4	13.0
72	9	24	scl	0.5	0.1	0.1	3.5	1.6	4.3	14.4
80	9	26	scl	0.4	0.1	0.1	3.5	1.6	4.3	16.0
88	11	29	sl	0.4	0.1	0.2	4.2	1.6	4.4	17.7
Mean	9.1	15.3		0.6	0.2	0.1	2.9	1.6	4.9	10.3

[a] c = clay (kaolinite), l = loam, s = sand.
[b] Dry-weight basis.

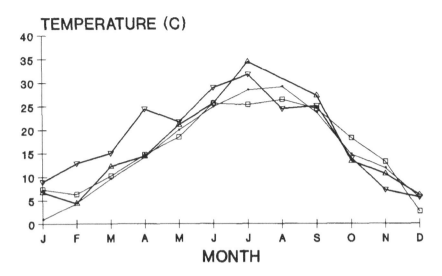

FIGURE 1 Mean monthly air temperature for 1988 (■), 1989 (□) and mean monthly soil temperature (7.6 cm depth) for 1988 (Δ) and 1989 (∇).

months of June to October) for 1988 and 1989 and the 10-year average. Water input on the lysimeters was approximately 25% above the 10-year average for the 1988 and 1989 seasons. Mean annual rainfall for the area is 107 cm[32] and annual runoff ranges from 3 to 18 cm.[33] Since the lysimeters were installed with a 3-cm lip, allowing for no runoff, the amount of water that infiltrated was also from 2 to 17% above normal.

Pan evaporation of water for the site for 1988 and 1989 generally followed the evapotranspiration losses for the region as shown in Figure 2B.

Water Flow Patterns in the Soil Core

Figure 3 depicts the concentration of tritium in the soil core after the soil has been (mathematically) unrolled and laid out in a 1.7-cm thick layer from the column wall to the center of the core, like unrolling a "jelly roll". The column was cut into three sections at the 25-, 50-, and 75-cm depths. The tritium and water front apparently moved down macro pores first, followed by micro pores, and became distributed in an icicle-like pattern at each soil depth. The maximum and most variable concentration of tritium was distributed in the 50-cm depth, but the distribution at the column wall was the same as that in the center of the core, at all depths, suggesting that water moved through the soil cores in an icicle-like front but in a relatively horizontal pattern. The greater variability was likely due to the presence of plinthite nodules and zones which diverted water movement somewhat. No wall-flow or core-flow of water was apparent.

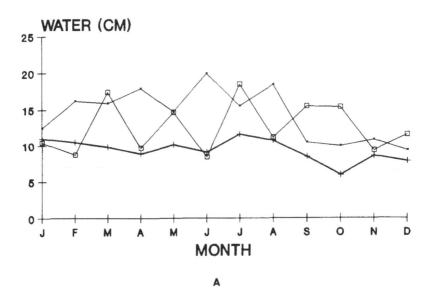

A

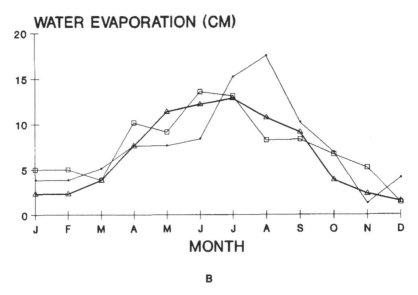

B

FIGURE 2 Mean monthly rainfall plus irrigation (A) for 1988 (■), and 1989 (□), and 10-year average rainfall (+), and mean monthly pan evaporation (B) for 1988 (■) and 1989 (□), and monthly evapotranspiration (B) from meadow grass at Raleigh, NC for 1961 (Δ) (5).

Longevity and Mobility of the Herbicides

Figure 4 depicts the mean distribution of ^{14}C in the 12 sections of the soil profiles from applications of ^{14}C-atrazine (Figure 4A), ^{14}C-metolachlor (Figure

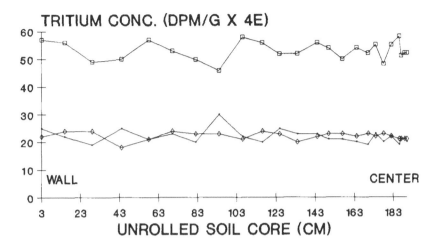

FIGURE 3 Mean tritium distribution at three depths [25, 50, 75 cm (■), 50 cm (□), and 75 cm (◇)] in 20-cm diameter × 90-cm long cores of undisturbed Dothan soil unrolled like a "jelly roll" from wall to center of core.

4B), and ^{14}C-primisulfuron (Figure 4C) at 30, 90, 180, and 365 DAT (days after treatment). From 30 to 35% of each of the herbicides was found in the 4-cm section and from 5 to 15% in the 11-cm section at 30 DAT. Less than 5% of the herbicides were found in the 19-cm section and lesser amounts were found in the deeper sections at 30 DAT. Less than 50% of each parent herbicide was found in the 0- to 11-cm sections and no parent herbicide was found below 27 cm. Since very small amounts of the herbicides were found in leachate, more than 50% of the compounds escaped as vapors or as $^{14}CO_2$.

At 90 DAT, ^{14}C from atrazine, metolachlor, and primisulfuron had decreased to 23, 19, and 14%, respectively, in the 4-cm section, to 8, 9, and 16%, respectively, in the 11-cm sections, and to 4% for each of the compounds in the 19-cm sections. Lesser amounts of ^{14}C were found in the soil profiles at 180 and 365 DAT and none of it was parent herbicide. Half-life (T–1/2) values computed from first-order equation plots of the three herbicides yielded values of 24, 38, and 30 days for atrazine, metolachlor, and primisulfuron, respectively.

Herbicide mobility (R_f) values calculated for each of the three herbicides in the soil profile at the four sampling times are shown in Table 3 and suggest that the order of mobility of the three herbicides was metolachlor > primisulfuron > atrazine, which had mean R_f values of 0.26, 0.16, and 0.11, respectively. Atrazine was probably the least mobile of the herbicides due to its weakly basic character and tendency to be ionically bound to organic and inorganic colloids in this acid soil.[7,34] Metolachlor was probably the most mobile compound due to its nonionic character and moderately high water solubility.[35] Primisulfuron was intermediate in its movement because, although it has weakly acidic properties, the parent compound has low water solubility, and the major metabolite is a sulfonamide derivative which is likely to have lower soil mobility than the parent.[36]

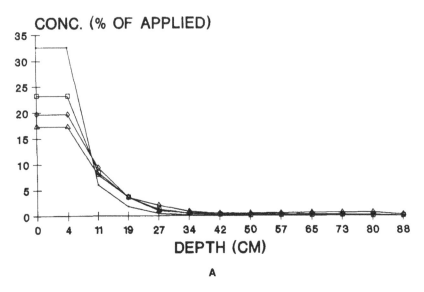

A

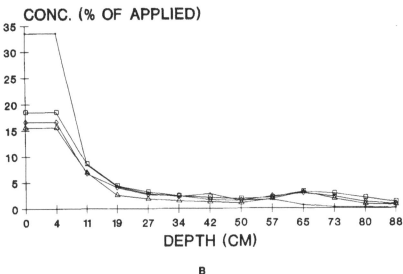

B

FIGURE 4 ^{14}C-distribution with depth in lysimeters treated with ^{14}C-atrazine (A), ^{14}C-metolachlor (B), and ^{14}C-primisulfuron (C) at 30 (■), 90 (□), 180 (◇), and 365 (Δ) days after treatment (1989, fallow, no leachate collectors).

Effect of Plant vs. Fallow

Figure 5 depicts the ^{14}C-metolachlor distribution in Dothan soil columns at 90 DAT for fallow (no plant) and soybean plant systems. The presence of the soybean plant reduced ^{14}C movement downward through the soil. Soil columns

CONC. (% OF APPLIED)

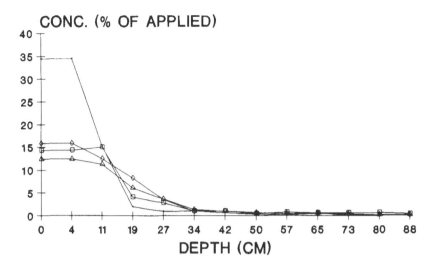

DEPTH (CM)

FIGURE 4C (continued)

Table 3. Calculated R_f Values for ^{14}C-Labeled Atrazine, Metolachlor, and Primisulfuron Distribution in 90-cm Long Undisturbed Columns of Dothan Loamy Sand (1989, Fallow, No Leachate Collectors)

Days after treatment	Calculated R_f values		
	Atrazine	Metolachlor	Primisulfuron
30	0.07	0.16	0.10
90	0.10	0.30	0.19
180	0.12	0.29	0.17
365	0.16	0.28	0.19
Mean	**0.11**	**0.26**	**0.16**

with soybean plants had 9% more ^{14}C in the upper two thirds of the column (53 vs. 45%) and 8% less ^{14}C in the lower one third of the column (4 vs. 12%) than columns with no plants. In addition, 5% of the ^{14}C was found in plant tissues (leaves and roots) and 99% less ^{14}C was found in the leachate of columns with plants than columns without plants (0.1 vs. 7.6%). Columns with plants also resulted in 68% less leachate (2.5 l vs. 8.0 l) and 34% lower mean soil moisture levels (7.8 vs. 11.8%) than columns without plants. Lastly, R_f values for columns with plants were 26% smaller than for columns without plants (0.23 vs. 0.31), suggesting that herbicide mobility was much slower through columns that had a soybean plant growing in them than through fallow soil columns. The difference in apparent herbicide mobility was caused primarily by the lower water movement through soil columns containing plants, due to the usage of the water by the plants since all of the columns received the same amount of input water. Total ^{14}C recovered in the two systems was similar, i.e., 62.3% for columns with soybean vs. 63.8% for fallow columns. Kilmer et al.[4]

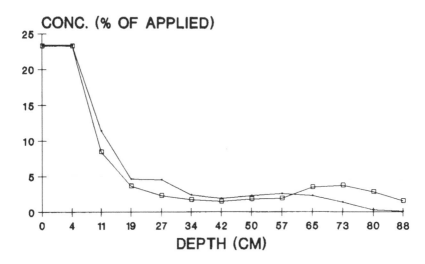

FIGURE 5   ^{14}C-distribution in lysimeters 90 DAT with ^{14}C-metolachlor in fallow (□) and soybean plant (■) systems (mean values for 1988 and 1989, with leachate collectors).

also reported that percolate from large 91-cm diameter by 112-cm deep cylindrical fallow lysimeters was up to 10 times that from lysimeters with corn plants growing in them.

Effects of Surface Cover

Figure 6 depicts the ^{14}C-metolachlor distribution in Dothan soil columns at 90 DAT for fallow, Bermuda sod, and wheat straw mulch systems. The Bermuda sod was more effective than soybean in reducing ^{14}C movement downward through the soil. There was 11% more ^{14}C found in the 4-cm section of the sodded soil columns than in the fallow columns (30 vs. 19%) and 18% less was found in the 11- to 88-cm sections of the sodded columns than the fallow columns (20 vs. 38%). There was 3% less ^{14}C found in the soil in the sodded column than in the fallow column and that 3% was found in the sod thatch itself. Total ^{14}C recovered in the sodded vs. fallow columns was similar, 52.3 vs. 52.1%, respectively. The mean R_f value for the sodded columns was 66% smaller than that for the fallow columns (0.20 vs. 0.30), also suggesting that the ^{14}C-metolachlor distribution downward in the soil was greatly reduced by the presence of the sod. Several investigators have reported that several pesticides that are applied to turf remain associated with the thatch and do not move deeply into the soil,[37,38] and metolachlor applied to turf is estimated to have a lower leaching potential than when applied to conventionally farmed systems.[31] It should be noted, however, that turf systems such as on golf courses and home lawns normally receive approximately twice as much water input, in the form of irrigation, as conventional or conservation-tillage crop-

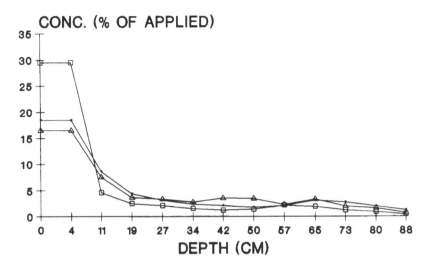

CONC. (% OF APPLIED)

DEPTH (CM)

FIGURE 6 ^{14}C-distribution in lysimeters 90 DAT with ^{14}C-metolachlor in fallow (■), Bermuda sod (□), and straw-mulched (△) systems (1989, no leachate collectors).

ping systems. This added water may greatly increase the mobility of metolachlor and other herbicides in turf situations.

In the wheat straw-mulched lysimeters (Figure 6), approximately 4% less ^{14}C-metolachlor was retained in the upper 19 cm of soil (28 vs. 32%) and approximately 4% more ^{14}C-metolachlor was found in the 27 to 66 cm depth (19 vs. 15%), as compared with fallow lysimeters, suggesting that the mulch increased downward movement of the herbicide. Mean R_f values were 7% higher for mulched lysimeters than for fallow lysimeters (0.32 vs. 0.30, respectively). Mean soil moisture levels for mulched vs. fallow lysimeters were quite similar, however, 11.6 vs. 11.7%, respectively, and 3.9% of the applied ^{14}C was found in the straw mulch itself. There was 4% more ^{14}C recovered in the mulch system than in the fallow system (56 vs. 52%, respectively), suggesting that less herbicide volatilized from the mulched columns. These columns were not equipped with leachate collectors, but a comparison of the mean soil moisture levels suggests that less leachate moved through the sodded and fallow columns than through the mulched columns.

Effect of Liming on Atrazine Mobility

Figure 7 depicts the ^{14}C-atrazine distribution in fallow soil columns at 90 DAT as influenced by liming. Liming the surface soil had the effect of reducing ^{14}C in the 4-cm section by 2% (21 vs. 23%) and increasing ^{14}C in the 11-cm section by 6% (14 vs. 8%) as compared with the unlimed treatment. There was 5% more ^{14}C also recovered in the lime soil compared with the unlimed soil (42 vs. 37%), suggesting that the lime increased the longevity of atrazine. A

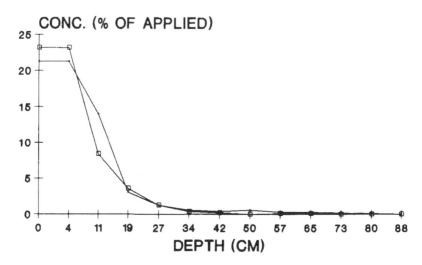

FIGURE 7 ^{14}C-distribution in lysimeters 90 DAT with ^{14}C-atrazine in limed (■) and unlimed (□) systems (1989, fallow, no leachate collectors).

computed R_f value for the limed soil was 30% higher than one computed for the unlimed soil (0.13 vs. 0.10), suggesting that the lime also increased the mobility of the herbicide. Soil pH values for the 0- to 7-cm section of the limed soil were 6.3 compared with 6.0 for the respective unlimed soil. For the 7- to 16-cm section, pH values were 5.8 and 5.6, respectively, and the mean pH values for the soil profiles were 4.9 and 4.8, respectively, suggesting that the lime had leached through the soil profile, but that the majority was retained in the surface horizons. It is likely that the pH effect on atrazine would be much more pronounced on calcareous soils where the pH of the entire subsoils would be in the range of 6.5 to 8.5. Other studies have shown that liming acid fields increased the bioactivity and longevity of atrazine.[7,40] These columns were not equipped with leachate collectors.

Effect of Leachate Collectors

Figure 8 depicts the ^{14}C-metolachlor distribution in fallow soil columns equipped with or without a leachate collector. There was 4% more ^{14}C recovered in the 4-cm soil sections of the columns equipped with a collector compared with the columns with no collector (22 vs. 18%) and 4% less ^{14}C was distributed in the 11- to 88-cm sections of the collector-equipped columns than the columns with no collector. There were 8 l of leachate containing 3% of the ^{14}C recovered from the columns equipped with collectors, so 3% more ^{14}C was also recovered from these columns, 54.7 vs. 52.3%, respectively. The mean R_f value for ^{14}C-metolachlor in the leachate collector-equipped columns was 7% smaller than for the columns with no collector (0.28 vs. 0.30), suggesting that the presence of the leachate collector may have slightly reduced the downward

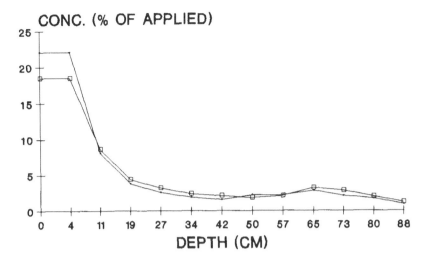

FIGURE 8 ^{14}C-distribution in lysimeters equipped with (■) and without (□) leachate collectors 90 DAT with ^{14}C-metolachlor (1989, fallow).

movement of the ^{14}C or reduced the upward movement of ^{14}C in capillary flow of water, possibly by interrupting capillaries at the 90-cm depth. It was observed that drops of water were hanging from the bottom of the soil column equipped with a leachate collector, and the drops had to grow to a reasonable size before they fell into the funnel.

Effect of Column Diameter

Figure 9 depicts the ^{14}C-metolachlor distribution in 10-, 15-, and 20-cm i.d. fallow soil columns at 90 DAT. [14]C distribution in the 15- and 20-cm i.d. columns was similar (52.0 vs. 52.3% recovered, respectively, and R_f values of 0.30 vs. 0.30, respectively). [14]C distribution in the 10-cm i.d. column was very different from that of the other 2 columns. There was 11% less ^{14}C recovered in the 10-cm i.d. soil column (41 vs. 52 and 52%, respectively), and the R_f value was 47% smaller (0.16 vs. 0.30 for the other two columns, respectively), suggesting that the ^{14}C was much less mobile in the 10-cm i.d. column than in the other two columns. It was noted that the soil level in the 10-cm i.d. column was also 5 cm lower than the surrounding soil, suggesting that the soil within the column had been compacted to some degree and this probably affected the rate of water and ^{14}C movement through the column.

SUMMARY

Tritium in percolating water through soil column field lysimeters moved in an icicle-like pattern at the moving front and moved horizontally with no

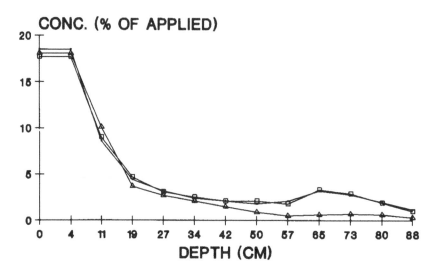

FIGURE 9   ^{14}C-distribution in 20-cm (■), 15-cm (□), and 10-cm (Δ) diameter soil column lysimeters 90 DAT with ^{14}C-metolachlor (1989, fallow, no leachate collectors).

apparent difference in mobility at the walls than at the soil core.

[14]C-distribution in the field lysimeters showed that metolachlor was more mobile than primisulfuron and atrazine, respectively. Differences were due primarily to differences in chemical and biological properties of the compounds and the high acidity of the soil.

The presence of a soybean plant growing in a lysimeter reduced the downward movement of ^{14}C-metolachlor by 26% in the soil, and by 99% in the leachate compared with fallow lysimeters. The plants also reduced the amount of leachate collected by 68% and the mean soil moisture content of the soil cores by 34% primarily because of the usage of water by the growing plants. Bermuda sod-covered lysimeters reduced ^{14}C-metolachlor in the soil by 33% compared with fallow lysimeters and wheat straw-mulched lysimeters increased herbicide mobility by 7% compared with fallow lysimeters.

Liming the soil increased ^{14}C-atrazine mobility in the soil by 30% and increased the amount recovered by 5%.

[14]C-metolachlor distribution in 15-cm and 20-cm i.d. columns were similar, but mobility in 10-cm i.d. columns was 47% slower. Leachate collectors reduced ^{14}C-metolachlor movement in the soil by 7% as compared with lysimeters with no leachate collectors attached.

ACKNOWLEDGMENTS

The authors acknowledge the Water Resources Research Institute and the U.S. Geological Survey (Grant No. 89-0496) and PTR Laboratory for supporting these studies, CIBA-GEIGY Corporation for providing materials, Len Swain, Wallace Baker, Henry Marshall, George Clark, Robert Lee, Mike Blumhorst, Jane McKinnon, Gail Mahnken, and Dr. C. T. Miller for technical assistance, and Dr. J. C. Wynne for cooperation and support.

LITERATURE CITED

1. Kohnke, H.; Dreibelbis, F. R.; and Davidson, I. M. A. *A Survey and Discussion of Lysimeters and a Bibliography on Their Construction and Performance;* Misc. Publ. No. 372; U.S. Dept. of Agric., Washington, DC, 1940.
2. Joffe, J. S. *Soil Sci.* 1932, 34, 123–143.
3. Musgrave, G. W. *Soil Sci.* 1935, 40, 391–401.
4. Kilmer, V. J.; Hays, O. E.; and Muckenhirn, R. J. *J. Am. Soc. Agron.* 1944, 36, 249–263.
5. Van Bavel, C. H. M. *Soil Sci. Soc. Am. Proc.* 1961, 25, 138–141.
6. Tackett, J. L.; Burnette, E.; and Fryrear, D. W. *Soil Sci. Soc. Am. Proc.* 1965, 29, 218–220.
7. Best, J. A. and Weber, J. B. *Weed Sci.* 1974, 22, 364–373.
8. Bowman, B. T. *J. Environ. Qual.* 1988, 17, 689–694.
9. Edwards, W. M. and Glass, B. L. *Bull. Environ. Contam. Toxicol.* 1971, 6, 81–84.
10. O'Conner, G. A.; Wierenga, P. J.; Cheng, H. H.; and Doxtader, K. G. *Soil Sci.* 1980, 130, 157–162.
11. Weber, J. B. In *Research Methods in Weed Science;* Wilkinson, R. E., Ed.; Southern Weed Science Society; POP Enterprises, Inc., Atlanta, GA 1972, pp. 145–160.
12. White, R. E.; Dyson, J. G.; Gerstl, Z.; and Yaron, B. *Soil Sci. Soc. Am. J.* 1986, 50, 277–283.
13. Witt, W. W. and Weber, J. B. *Weed Sci.* 1975, 23, 302–307.
14. McMahon, M. A. and Thomas, G. W. *Soil Sci. Soc. Am. Proc.* 1974, 38, 727–732.
15. White, R. E.; Thomas, G. W.; and Smith, M. S. *J. Soil Sci.* 1984, 35, 159–168.
16. Cassel, D. K.; Krueger, T. H.; Schroer, F. W.; and Norum, E. B. *Soil Sci. Soc. Am. Proc.* 1974, 38, 36–40.
17. Richter, G. and Jury, W. A. *Soil Sci. Soc. Am. J.* 1986, 50, 863–868.
18. Saffigna, P. G.; Keeney, D. R.; and Tanner, C. B. *Soil Sci. Soc. Am. J.* 1977, 41, 478–482.
19. Jordan, C. F. *Soil Sci.* 1968, 105, 81–86.
20. Kubiak, R.; Fuhr, F.; Mittelstaedt, W.; Hansper, M.; and Steffens, W. *Weed Sci.* 1988, 36, 514–518.
21. Weber, J. B. Leaching Column and Method of Use; U.S. Patent No. 4,166,720, U.S. Patent Office, Washington, DC, 1979.
22. Lawrence, L. J.; Ruzo, L. O.; and Olsen, G. L.; Method and Apparatus for Conducting Field Dissipation and Leaching Studies; U.S. Patent No. 5,009,112, U.S. Patent Office, Washington, DC, 1991.
23. Mehlich, A. *Commun. Soil Sci. Plant Anal.* 1984, 15, 1417–1422.
24. Walkley, A. and Black, J. A. *Soil Sci.* 1934, 37, 29–38.

25. Nelson, D. W. and Sommers, L. E. In *Methods of Soil Analysis;* Page, A. L., Ed.; Agronomy No. 9; Agronomy Society of America, Inc., Madison, WI, 1982, Part 2; pp. 539–580.

26. Gee, G. W. and Bauder, J. W. In *Methods of Soil Analysis;* Klute, A., Ed.; Agronomy No. 9; American Society of Agronomy, Inc., Madison, WI, 1986, Part 1; pp. 404–409.

27. Mehlich, A. *Commun. Soil Sci. Plant Anal.* 1984, 15, 1409–1416.

28. Blake, G. R. and Hartge, K. H. In *Methods of Soil Analysis;* Klute, A., Ed.; Agronomy No. 9; Agronomy Society of America, Inc., Madison, WI, 1986, Part 1; pp. 363–375.

29. *The Triazine Herbicides;* Gunther, F. A., Ed.; Springer-Verlag, New York, NY, 1970.

30. *Herbicide Handbook;* Weed Science Society of America, Champaign, IL, 1989.

31. Weber, J. B. and Warren, R. L. *Proc. Northeastern Weed Sci. Soc.* 1993, 47, 144–157.

32. Van der Leeden, F.; Troise, F. L.; and Todd, D. K. *The Water Encyclopedia;* 2nd. ed.; Lewis Publishers, Inc., Chelsea, MI, 1990.

33. *Control of Water Pollution from Crop Land.* Vol. I. A Manual for Guideline Development; U.S. Environmental Protection Agency, Washington, DC, 1975.

34. Weber, J. B. In *Fate of Organic Pesticides in the Aquatic Environment;* Gould, R. E., Ed.; Advances in Chemistry Series No. 111; American Chemical Society, Washington, DC, 1972, pp. 55–120.

35. Peter, C. J. and Weber, J. B. *Weed Sci.* 1985, 33, 874–881.

36. Beyer, E. M.; Duffy, M. J.; Hay, J. V.; and Schlater, D. D. In *Herbicides; Chemistry, Degradation, and Mode of Action;* Kearney, P. C. and Kaufman, D. D.; Eds.; Marcel Dekker, Inc., New York, NY, 1988, Vol. 3; pp. 117–189.

37. Braham, B. E. and Webner, D. J. *Agron. J.* 1985, 77, 101–104.

38. Sears, M. K.; Bowhey, C.; Braun, H.; and Stevenson, G. R. *Pestic. Sci.* 1987, 20, 223–231.

39. Weber, J. B. *North Carolina Turfgrass* 1991, 9, 24–29.

40. Lowder, S. W. and Weber, J. B. *Weed Sci.* 1982, 30, 273–280.

CHAPTER 4

Studies on Pesticide Mobility:
Laboratory vs. Field

R. F. Turco and E. J. Kladivko

The transport of agricultural chemicals through soils and into ground water is a national concern. Data released in 1990 from the EPA national survey of pesticides in drinking water indicate some 446,000 of the nation's rural domestic wells may contain at least one pesticide. Although concentrations are usually low, significant occurrences of pesticides have been found and attributed to non-point sources of contamination. Current understanding of the basic mechanisms of pesticide transport and transformations in the vadose zone is not sufficient to quantitatively predict these occurrences and in some cases the lack of occurrence of pesticides in water. In particular, the presence of small amounts of pesticides much deeper in soil profiles than predicted by standard water and chemical flow models is problematic. This "preferential flow" or fast flow of some fraction of the applied chemical may be important as a significant amount of chemical may reach ground water before it has had time to be attenuated by the environment.

The purpose in investigating the environmental fate of agricultural chemicals is to understand the potential risk they may pose to humans and the environment. How a pesticide behaves in the environment is governed by retention, transformation, and transport processes. Any effort to understand the fate of a chemical in the environment is in fact an effort to define the role of these three processes for a given chemical. Hence, assessments of the fate of a pesticide in the environment must include an estimation of all processes that

0-87371-926-3/94/$0.00+$.50
© 1994 by CRC Press, Inc.

63

could potentially affect that compound.[1] Proper assessments can only be made with both field and laboratory data.

The underlying assumption inherent in many previous discussions of transport of solutes in field soils was a belief that models developed to describe the transport of water and chemicals in laboratory columns of uniformly packed porous media do in fact represent the field process.[2,3] This bias follows some 50 years of research that has utilized packed soil columns and water soluble tracers such as bromide as well as sorbing pesticides.[4] These approaches have failed to describe lateral variability in transport[5] or the rapid transport of water through soil macropores first described by Lawes et al. in 1881[6] but more recently linked to pesticide movement to ground water.[2] Moreover, most environmental fate models use laboratory-derived sorption coefficients to describe chemical binding,[7] assume microbial degradation follows only first order kinetics, and discount the importance of the rhizosphere as an environment for pesticide reactions.

The purpose of this article is to point out major areas where we should be considering the impact of actual field conditions on inferences derived from laboratory data. This is particularly true for the prediction of chemical leaching in the field, the magnitude of which is a reflection of chemical, physical, and biological processes occurring in a complex soil structure, often under non-steady-state conditions. We also point out the strengths of tile-line lysimeters as an environmental assessment device for the prediction of chemical migration in soil. We begin by discussing chemical attenuation processes that are generally evaluated at a relatively small scale (sorption/desorption, biodegradation), and then we discuss transport processes at a slightly larger scale.

SORPTION/DESORPTION

Laboratory studies provide a basis for understanding the fate of agricultural chemicals in the field. For a given agricultural chemical, development of a sorption coefficient or K_f value is a common starting point to describe its behavior. This sorption value is derived from fitting a power function known as the Freundlich equation:

$$S = K_f \, C^{1/n}$$

in which S is the sorbed concentration (μmol kg^{-1}), C is the equilibrium solution concentration (μmol L^{-1}), and K_f and 1/n are empirical constants. This equation has been used to describe the sorption behavior of most agricultural chemicals as well as other common soil elements. When n = 1 a linear equation results, and K_f becomes K_d, the linear distribution coefficient (l kg^{-1}).[8] If 1/n is greater or less than 1, the sorption isotherm takes on a concave or convex shape, respectively.[9] The significance of considering the deviations of the 1/n value from 1 has been considered by others.[7,10] It is suggested that for pesticides $1/n$ tends to less than 1 with a mean value of 0.87.[7]

In general, K_f is derived using the customary 24-h batch equilibration procedure. The batch procedure produces repeatable and widely used estimations. However, does the batch process actually represent the partition mechanisms at work in the field? Bilkert and Rao[11] found that K_d values were incorrect in calculation of the movement of two nematicides. They concluded that because the system was in a dynamic non-equilibrium state, the use of an equilibrium value such as K_d in a pesticide transport model tended to overestimate both sorption and desorption. Pignatello and Huang[12] reported that for times up to 15 months following application to soil, large amounts of atrazine and (metolachlor) are maintained in a slowly reversible form. They report an apparent distribution constant K_{app} that is derived from consideration of the bound forms of the pesticide.

The use of a distribution coefficient such as K_d assumes that both sides of the sorption process are instantaneous. Hence, an overestimate of the amount of pesticide released in a desorption step may result in an overestimation of the potential amount available for leaching. Moreover, because the two sides of the sorption process are not instantaneous, conclusions about the timing of pesticide movement, when based on K_d, may also be flawed. Approaches to understanding the kinetics of the process have centered on a number of models. These models have attempted to understand how pesticides distribute themselves in soil.

While artifacts and methods problems may play a role in the observation of hysteresis, the general conclusion is that hysteresis is real and a significant factor in describing the fate of pesticides in soil. Pignatello,[13] reporting on the work of Karickhoff,[14] Karickhoff and Morris[15] and Selim et al.,[16] provided an approach to understanding the process. This model, sometimes referred to as a two-box model, implies that two types of chemical sorption site exist in soil. They define a system which has a solution phase (C) and sorption sites that are either labile (S_1) or non-labile (S_2). The relationship between the phases is defined by:

$$C \xleftrightarrow{X_1 K_d} S_1 \xleftrightarrow{k} S_2$$

where X_1 is the fraction of sorbed chemical in the S_1 state, K_d is the distribution coefficient and k is the first order rate constant for movement in and out of the S_2 phase. The S_1 phase or sites are considered to be rapid in their release of materials. It should be noted that while the S_2 are slower to release materials, they will at some point release the organic compound. Although the model indicates that release of the materials from S_2 goes through S_1, the potential for a direct interaction between S_2 into the solution phase is also possible.[15] This two-site approach has been used to explain the fate of picloram in soil.[13] Following sorption, a significant decrease in the amount of the chemical released over time was observed; the two sites were described as either having desorption times of 5 h (S_1) or 300 h (S_2). The sites were proposed to be on soil surfaces (S_1) and internal (S_2) to the soil surface.

Another approach to considering the sorption of organics is with diffusion kinetics. It is clear from the work of Wu and Gschwend[17] that the rate at which an organic is taken up and released is inversely dependent on the diffusion path-length and tortuosity. Hence, for a given soil the size and shape of the aggregates will have a major effect on the penetration of the organic from the pore water into the soil. The fate of the soil fumigant EDB has been explained with a diffusion model. EDB is water soluble, has a small K_d and is considered to be biodegradable. However, EDB has been detected in soils as much as 19 years after the last application.[18] It is theorized that EDB becomes trapped in microsites and is protected from leaching or microbial degradation.[18] Laboratory sorption studies utilizing uniformly sieved soils and short time periods will overlook this process.

It follows from either approach that the fate of soil-applied chemicals is a function of types and arrangement of sorption sites. Further, because soil is not uniform, the mixture of available sorption sites will have a major effect on considerations of pesticide movement. Hence, use of uniformly packed soil columns will alter the sorption kinetics typically found in field conditions, as well as the sorption mechanism. Hance[19] has indicated that desorption equilibrium may take longer to reach than adsorption equilibrium. This slower equilibrium may limit the amount of herbicide that is present in solution. Hamaker and Thompson[10] stated that once a material is bound it begins to age and over time becomes more difficult to remove. Wauchope and Meyers[20] further define this idea of aged residues by using a sequential equilibria model, similar to the two-box model,[13] to describe equilibria occurring on surfaces.

While their work was confined to conditions involving only a few minutes (typical of runoff or stream conditions), it can be seen that given enough time a significant portion of the applied materials will be transferred from the labile rapidly reversible sites, to the more restricted sites others have shown. This type of transfer is not always coupled to a modification in the chemical structure.[21] Spillner et al.[22] found that 48 to 50% of the applied ^{14}C-fenitrothion was moved to non-degradable (slowly released) sites in 40 to 50 days post application. Subbarao and Alexander[23] have shown that the desorption rate of organics from clays into lake water is the rate limiting step in degradation.

DEGRADATION

Biological degradation, either complete mineralization or partial decay to form metabolites, is the critical process controlling the ultimate fate of chemicals in soil and water. Accurate prediction of the degradation behavior is needed to allow calculation of environmental fate. It is well known from both field and laboratory studies that soil microorganisms can attack many pesticides. The size of the microbial population that can degrade a particular compound is difficult to estimate. For too long researchers have used plate counts as a way of estimating degradation potential of field and laboratory soils. This procedure tends to select for those organisms that grow well on agar and not those that function in soil. However, with selective in-soil biomass labeling procedures or more advanced gene probe technology, we may be able to make good estimations of population size. Selective labeling has been used to estimate population size in carbofuran-enhanced soils.[24] Clearly the better approach is made by making population estimations in the field. Laboratory studies tend to simplify the environment in which bacteria reside, and the major environmental factors influencing pesticide degradation (moisture, temperature, aeration, and pH) are optimized.

The primary problem facing a microorganism trying to use a soil-applied material is one of contact between the two. Contact may be achieved by either direct physical contact between herbicides adsorbed to soil surfaces and the cells, or by the movement of herbicide in solution to cells. Direct contact between adsorbed herbicide and microorganisms is limited by the size of the degrading population. Contact by diffusion in soil-water is controlled by population size, the fraction of herbicide in solution, and the physical limitations to diffusion of the soil-water. Most pesticide degradation studies are conducted in the laboratory under constant and ideal moisture contents, and many have been conducted in slurries. Provided aeration is adequate, these studies will tend to overestimate pesticide degradation that could occur in the field because the studies generate a good microbe-pesticide contact. Gerst and Yaron[25] show that as soil moisture is decreased from 60% water-filled pores (WFP), degradation rate of a herbicide is radically decreased. This has also been reported by other workers.[26,27] The interaction of moisture and pesticide degradation limits the usefulness of any model that ignores the interaction by assuming a constant half-life regardless of moisture content. Walker[27] proposed to overcome the limitation by using a relationship where half-life depends on moisture content, $T = aM^{-b}$ where T is the half-life, M the moisture content, and a and b are fitted constants.

Adsorption to surfaces has been shown to limit the degradation of herbicides in soil. Allen and Walker[28] found a strong negative correlation between the amount of herbicide adsorption and its degradation. They concluded that in the case of metolachlor, a statistically significant relation existed between microbial respiration and the amount of the herbicide in solution. This implies that, for at least some herbicides, the degradation will occur slowly where the population is inactive or the concentration in solution is low. Weber and Coble[29] have shown that the type of surface onto which herbicides were adsorbed may impact their degradation. Montmorillonite clays significantly reduced the degradation of diquat as compared to the same amount of kaolinite. Work by Ogram et al.[30] has indicated that 2,4-D, once adsorbed to soil, was completely protected from both solution and surface resident bacteria. The bacteria metabolized 2,4-D intercellularly.

In general, sorption of organics is controlled by the presence of soil organic matter. Organic matter acts upon herbicides and other pesticides in several ways. Organic matter, including the various unstabilized nonhumic materials produced from the decomposition of plant residues, affects both the adsorption of herbicides to soil and the microbial community that degrades herbicides. The tendency for organic matter to reduce herbicide availability through adsorption will not result in complete removal of the chemical. Rather, it appears that soil organic matter may act to modulate the amount in solution, as it supplies different types of sorption sites and diffusion path lengths.

Binding of herbicides to soluble soil organic materials can result in formation of soluble complexes which may not be sorbed. Researchers have demonstrated interactions between pesticides, such as triazines, and soluble organic materials, especially humic and fulvic acids.[31-33] It is thought that partitioning into a water soluble organic phase may explain discrepancies in predicted pesticide transport and soil/water partitioning in soils.[34] These types of interactions could significantly change the fate of pesticides in soil and may be critical to include in modeling efforts. The potential for the microbial biomass to remove a bound material from soil has been reported. Mac Rae[35] has shown that release of ^{14}C-residues held in soil is increased by the addition of substrates such as wheat straw or chemicals with structures similar to those first applied. Further, it was shown that the amount released was different in different soils. Other studies have demonstrated the potential for released residues to affect plants.[36,37]

Although authors have indicated that the rhizosphere may have a major impact on the fate of pesticides in soil,[38,39] little experimental work has been published. Root excretion and deposition of carbon around the roots allows for the development of a distinctly different region of microbial activity compared to the bulk soil. Compared to bulk soil, the rhizosphere can be enriched anywhere from 2 to 20 times in the number of organisms.[40] To a large degree the environmental fate of most pesticides is a function of the microbial biomass.[41] Hence, a stimulation in microbial number may increase the potential for degradation. However, the input of plant carbon could slow degradation, as a population of bacteria that prefers the plant material to the pesticide could develop.

Estimations of the rhizosphere effect are often drawn from studies where glucose is added to soil along with a pesticide. The carbon addition is thought to simulate the root exudates typical of the rhizosphere. Pettygrove and Naylor[42] reported an increase in mineralization of metribuzin following glucose addition. Similar results have been observed for atrazine and diuron degradation in glucose-amended soils.[43] In contrast, the addition of glucose decreased the rate of dalapon degradation.[44] Competition for other nutrients may be a critical factor in whether increased carbon will result in increased degradation.

In the actual rhizosphere, herbicide is delivered by mass flow as part of the plants transpiration stream. Hsu and Bartha[45] compared degradation rates of diazinon and parathion in bulk and rhizosphere soils. They found 8 and 10% higher rates for rhizosphere mineralization of diazinion and parathion, respectively. In contrast, 2,4-D degradation has been shown to be the same in the rhizosphere as in the bulk soil.[46] However, the buildup of nonextractable soil bound materials was increased under rhizosphere conditions.

Some herbicide passes through the rhizosphere and enters the root. Corn has been reported to take up about 20% of the applied atrazine; most of this atrazine is translocated to the shoots and detoxified by glutathione conjugation. These metabolites could then be returned to the rhizosphere or remain in the plant residue. These latter metabolites would be introduced to soil following plant harvest. Cheng et al.[47] hypothesized that plant metabolites of methabenzthiozuton (MBT) are more easily degraded than MBT itself.

The rhizosphere is known to be enriched with root-derived organic materials, including exudates, secretions, mucilages, mucigel, and lysates.[48] Estimates in non-sterile systems[50] indicate that as much as 20% of material assimilated by the plant could be lost as root exudates. However, since losses to the rhizosphere depend on many factors and since collection and analysis is difficult, it is hard to determine exact amounts of C released. Of particular interest is the fact that a substantial fraction of plant-deposited materials in the rhizosphere are high molecular weight compounds of the type shown to interact with pesticides.[50,51]

Presence of these organic compounds in the rhizosphere could increase pesticide adsorption and sorption-catalyzed hydrolysis of s-triazines, as well as other chemicals. Adsorption to organic matter may protect the herbicide from degradation. However, adsorption sites can also catalyze non-biological degradation such that adsorption can sometimes increase degradation rates.[38] For atrazine, reported effects of organic matter concentration are variable, and apparently depend both on bioavailability after adsorption and the extent of sorption-catalyzed reactions. Hurle[52] reported increased persistence of atrazine when straw ash was added to soil; Moyer et al.[53] reported a decrease in degradation for charcoal-amended soils. Recent work demonstrates a correlation between adsorption of imazaquin and concentration of organic matter.[54] These results suggest that simple CO_2 evolution from mineralization of herbicides in the rhizosphere may not provide the complete picture of degradation. The rich region around the roots may lead to cooxidation conversion to metabolites and bound residues and uptake into the roots.

TRANSPORT

Transport of chemicals through soils has been studied at both the laboratory and field scales. Until recently, most laboratory studies used columns of sieved, repacked soil to study movement of pesticides under steady-state water flow conditions. The chemical breakthrough curves under these conditions can usually be well described by classical convection-dispersion equations (CDE).[55] Although it was known that these idealized soil columns were rough simplifications of transport in the field, the degree of oversimplification of transport was not fully appreciated until recently. The occurrence of small amounts of pesticides much deeper in soil profiles than expected, or their arrival at a water table much faster than predicted, suggests that laboratory studies and CDEs are not sufficient to describe field-scale transport. This "preferential flow" or fast flow of some fraction of the applied chemical may be important if a significant amount of chemical reaches ground water before it has had time to be attenuated by the environment.

There is a growing body of evidence that preferential flow contributes to field-scale chemical transport[34,56-58] and is not adequately described by standard CDEs. Newer approaches to describing transport in structured soils include two-site sorption or two-region transport models[59,61] and stochastic approaches;[62,63] these approaches are reviewed by van Genuchten and Jury,[55] and Jury and Ghodrati.[64]

Although preferential flow of both mobile and adsorbing solutes has been observed in studies in unsaturated soils or shallow saturated zones, the relative importance of this phenomenon under different soil conditions is poorly understood. Jury et al.[34] reported in a field study on a loamy sand that approximately 20% of the applied mass of napropamide migrated far below the maximum depth predicted to be reached if it had undergone equilibrium adsorption. Similar amounts of preferential flow by other strongly adsorbing chemicals were reported in other studies on the same field by Clendening[65] and Ghodrati.[66] In all cases, the maximum depth reached by the sorbing chemicals was comparable to that reached by a mobile tracer added simultaneously, indicating that little or no adsorption was occurring in the preferential flow region. Other studies conducted on different soils have also shown evidence of incomplete sorption in preferential flow pathways.[57,67]

The occurrence of preferential flow in field soils is one of the major reasons why laboratory studies are of limited use for describing field-scale transport processes. These packed column studies may be useful in characterizing the fraction of an applied chemical that moves primarily through the soil matrix, but will not describe the interaction between matrix flow and preferential flow.

In order to include preferential flow processes in transport studies, experiments are conducted either at the field-scale or in the laboratory using intact soil columns sampled from the field. Large soil columns have been used to study the effect of water application rates, ponded vs. non-ponded conditions, tillage treatments, initial soil water contents, and transient water flow conditions on chemical transport including preferential flow.[68-73] These studies have provided useful data on the preferential flow process under controlled conditions. One of the limitations of these large column studies is that the lower boundary of the excavated column is not under the same soil water potential conditions as when it is connected to the underlying soil in the field. The researcher must choose whether to impose gravity flow conditions or to apply some tension at this lower boundary, and the choice of this boundary condition can affect the flow processes studied. Another limitation is that some macropores that are important in the field may terminate at the column wall, while other macropores that are not continuous in the field may be continuous through the limited length of the column sample.

Field scale preferential flow is difficult to study quantitatively, in part due to its spatially variable nature. Soil cores and suction cups have provided useful information but are limited by their small size compared with the presumed scale of the process. It is also impossible to take the prohibitive number of samples required to adequately describe the spatial variability in the field. Richard and Steenhuis[74] summarized many of the problems with using multiple small soil samples for trying to describe field-scale transport. They argued that tile drains might fulfill some of the requirements for the "ideal" measurement instrument for field-scale, as described by Cushman.[75,76] Because an important thing to know about preferential flow is its order of magnitude importance on a field-scale, it is appropriate to use an instrument that integrates over this same scale. Tile drains collect water and chemicals from a relatively well-defined and large area of a field and thus could serve as a useful research tool. Tile drains are also readily available at many research stations and production fields, and their outlets can be instrumented for relatively modest cost.

Tile drains have been used in the past to assess the impact of agricultural management practices on surface and ground water quality.[77-80] Hallberg et al.[79] provided a good review of these studies and concluded that tile drains can be useful tools for assessing the impact of agricultural management systems on ground water quality. Until recently, the methods employed in most of these studies were not designed to study preferential flow per se, although inferences about preferential flow can be made from some of the data. For example, Bottcher et al.[81] found small amounts of two chemicals with different sorption coefficients in drain outflow within 4 days of application. More recently, investigators have specifically used tiles to study preferential flow. Van Ommen[82] applied a non-sorbing tracer uniformly across a field of loamy sand in the Netherlands, and they observed some early breakthrough of the tracer. Richard and Steenhuis[74] applied a tracer in a 3.5-m wide strip offset 2 m from the tile line, and they also observed early breakthrough of the tracer. Our estimates from their data suggest that roughly 20% of the tracer arrived at the tile sooner than predicted. Everts et al.[83] irrigated a silt loam soil with a solution of two non-sorbed tracers and two differently sorbed tracers, and they detected all four tracers in tile outflow within one hour after the irrigation commenced.

In an ongoing field study in southeastern Indiana, we have been measuring pesticide and nitrate concentrations and total mass losses in a tile drain spacing experiment.[84] The soil is poorly structured, low in organic matter, and slowly

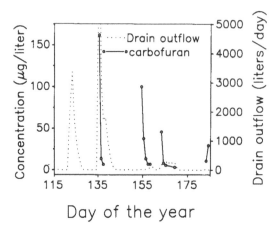

FIGURE 1   Subsurface drainflow and carbofuran concentrations from the 5-m spacing, Block 2, from Day 115 to 185, 1985. Carbofuran application occurred on Day 112.

permeable, and would not be expected to exhibit significant preferential flow. However, four pesticides with different sorption coefficients all arrived at the tile drain at the same time, less than 3 weeks after pesticide application and with less than 2-cm net water drainage from the soil. Since 1 pore volume of this soil equals about 30 cm, the early breakthrough suggests that less than 7% of the pore volume is effective in rapid transport. Although all four chemicals arrived at the same time, the rank-order of mass losses of chemicals corresponded to the rank-order of their sorption coefficients (K_d). Almost all the pesticide that was lost in tile outflow arrived during the early breakthrough period, again suggesting that the preferential flow component of transport may be the most critical for many agricultural chemicals.

Pesticide transport through the soil and into the drains appeared to be flow event-driven, and therefore cannot be described as a simple function of time or net drainflow volume. Carbofuran concentrations were high at the beginning of each new drainflow event (Figure 1), even when the new drainflow events were very small (see Day 154 in Figure 1, for example). Concentrations decreased rapidly as the flow event continued. Atrazine and cyanazine showed the same general response, but concentrations were much lower and more variable. There appeared to be some attenuation of the peak concentrations with each successive drainflow event. Samples for the first drainflow event (Day 122) after planting were not available; so, we do not know if chemicals had reached the subsurface drain by that time.

We proposed that the explanation for the observed behavior is nonequilibrium sorption/desorption in the preferential flow paths. At the start of a flow event, pesticide in the existing soil solution is flushed rapidly through large pores and into the drain. Desorption is not rapid enough to maintain an equilibrium solution concentration in new rain water; so, continued water flow through

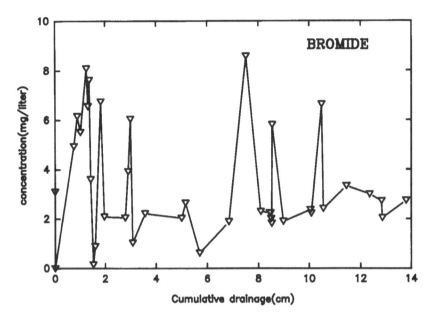

FIGURE 2 Bromide concentrations in subsurface drainflow from the 20-m spacing, Block 2, as a function of cumulative drainflow after bromide application.

those pores contains much lower concentrations. When drainage ceases, desorption supplies chemicals to the water in the large pores, and another rainfall would cause a new flow event to contain a high initial concentration of pesticides. In contrast, Hallberg et al.[79] suggested that the initial part of the flow event is dominated by drainage from large pores, in which water and chemicals have a short residence time. Therefore concentrations in the outflow are high. As the drainage slows down, most of the water being delivered to the drains is from smaller pores, where the chemical is retarded in the soil profile for longer times, and therefore concentrations are much lower. However, in our case, with less than 2-cm total net drainage from the soil, it is unlikely that any pesticide traveling through the matrix would have already reached the tile drain.

In November 1989 we spray-applied bromide and two differently sorbed pesticides to a 3-m-wide strip offset 1.2 m from the tile. Bromide (Figure 2) and both pesticides were detected in the first outflow after application, with less than 1-cm net drainage.[85] The data strongly support the hypothesis that preferential flow is occurring in the bulk of the field and not just in the disturbed soil of the former trench above the tile. About 6% of the applied tracer has been removed with 15-cm drainage water during the winter and early spring after application, suggesting a minimum of 6% of applied chemicals will move through preferential flow pathways in this soil.

Tile drains are a useful tool for studying chemical transport processes on a field-scale, but as with all methods, there are also some limitations. Tile drains flow only when the water table is at or above the depth of the tile. Thus, tile drains will not provide any water samples during those times of the year when the water table is below the tile. In many agricultural fields, tiles drain a seasonal high water table, and generally do not flow in mid-summer or early autumn. However, these are also the times of year when net downward movement of water and therefore chemicals would be small due to upward flow of water to the plant.

The installation of tile drains also involves digging a trench through the field and then backfilling above the tile, often with the original soil. This backfilled trench is initially loose and may provide a rapid flow path for water and chemicals. With time the trench settles, and the amount of preferential flow through the trench would presumably decrease. It might also be argued that the backfilled trench, with its disturbed soil, may behave more like a packed column, with large channels obliterated and flow more uniform throughout. More study is needed to clarify the influence of the trench on water and chemical flow, with time after tile installation. In any case, tile drains are a "fact of life" in many agricultural fields,[79] and as such provide estimates of materials leached below the rootzone in tiled fields. For studies of preferential flow per se, a technique to avoid the potential problem of the old trench is to apply a chemical in a narrow strip offset from the tile.[74,95]

The two-dimensional flow regime of water moving to tile drains complicates the interpretation of outflow concentrations.[86,87] Chemical concentrations found in tile outflow integrate the result of recent chemical applications near the tile and applications from several to many years previously at further distances from the tile. The spread of travel times through the saturated zone resulting from the different streamline distances can be estimated.[86,88]

In most cases tile lines do not collect all of the water that leaches below the rootzone, and some is lost to deep seepage. Thus tile drains cannot be used to obtain a complete water and chemical balance as is possible with monolith or weighing lysimeters. However, the quality of the water collected in the tiles reflects the quality of water lost as deep seepage.[79]

Tile lines drain a relatively well-defined area of the field (from midplane to midplane), if they are installed in a systematic pattern. Studies using tile drains usually incorporate border tiles, perimeter drains, or subsurface impermeable barriers around the plot, to further define the drainage area of each tile. Studies in Minnesota and Ohio have used plastic film in vertical trenches around the plots to prevent subsurface lateral movement of water into or out of the plots. Studies in England used perimeter drains of gravel and tile lines. In Indiana and Iowa, border tiles were used between each sampled tile line.

Tile drains have been shown to be useful for assessing overall chemical losses from the rootzone, and more recently have been used to study preferential flow processes. The abundance of tile-drained fields in many agricultural regions may give us a unique opportunity to measure field-scale chemical transport on a variety of soils and crop production systems.

SUMMARY

The environmental fate of a pesticide depends on a complex interaction of many processes. Laboratory studies of pesticide sorption, degradation, and transport have elucidated many of the basic principles involved in determining pesticide fate in the field. Pesticide sorption at different types of sites in the soil may be greatly affected by the range of aggregate and pore sizes in a natural field soil. Degradation rates are affected by microbial population size and degree of contact between organisms and pesticides, both of which can be affected by soil structure. Degradation rates may also be very different in the rhizosphere than in the bulk soil. Transport of pesticides is greatly affected by the presence of large structural cracks and channels in a natural field soil. "Preferential flow", or the rapid transport of small amounts of chemicals through the soil, is due in part to the presence of these structural features.

Improving our knowledge of pesticide fate in the environment requires a creative mix of laboratory and field research. One field research technique that is becoming more widely used is tile drain lysimeters. Subsurface tile drains integrate water and chemical flow on a field-scale basis and may be particularly helpful for studying preferential flow. When linked with studies of pesticide degradation in the rhizosphere and sorption/desorption in structured soils, a better understanding of pesticide fate in the environment will be achieved.

ACKNOWLEDGMENT

Paper No. 13,291 of the Purdue University Agricultural Experiment Station Series.

LITERATURE CITED

1. Cheng, H. H. "Pesticides in the soil environment — An overview"; Pesticides in the soil environment, H. H. Cheng, Ed., Soil Science Society of America, Madison WI.
2. Thomas, G. W. and R. E. Phillips. 1979. Consequences of water movement in macropores. *J. Environmental Quality* 8:149–152.
3. Jury, W. A. 1983. Chemical transport modeling: Current approaches and unresolved problems. pp. 49–64. In *Chemical Mobility and Reactivity in Soil Systems.* SSSA Special Publication 11.
4. Bodman, G. B. and E. A. Colman. 1943. Moisture and energy conditions during downward entry of water into soils. *Soil Sci. Soc. Am. Proc.* 8:116–122.
5. Rose, C. W., F. W. Chichester, J. R. Williams, and J. T. Ritchie. 1982. A contribution to simplified models for solute transport. *J. Environmental Qual.* 11:146–150.

6. Lawes, J. B., J. H. Gilbert, and R. Warington. 1881. On the amount and composition of the rain and drainage waters collected at Rothamsted. William Clowes and Sons, LTD. London.

7. Rao, P. S. C. and J. M. Davidson. 1980. Estimation of pesticide retention and transformation parameters required in non-point source pollution models. p. 23–67. In *Environmental Impact of Nonpoint Source Pollution*, M. R. Overcash and J. M. Davidson (ed.), Ann Arbor Sci. Publ. Ann Arbor, MI.

8. Shea, P. J. 1989. Role of humified organic matter in herbicide adsorption. *Weed Technology* 3:190–197.

9. Giles, C. H., T. H. MacEwan, S. N. Nakhwa, and O. Smith. 1960. Studies in adsorption. Part XI. A system of classification on solution adsorption isotherms, its use in diagnosis of adsorption mechanisms and in measurement of specific surface area of solids. *J. Chem. Soc.* 1960:3973–3993.

10. Hamaker, J. W. and J. M. Thompson. 1972. Adsorption. p. 49–143. In *Organic Chemicals in the Soil Environment*, C. A. I. Goring and J. W. Hamaker (ed.), Vol. 1. Marcel Dekker, New York.

11. Bilkert, J. N. and P. S. C. Rao. 1985. Sorption and leaching of three nonfumigant nematicides in soils. *J. Environ. Sci. Health* 320:1–26.

12. Pignatello, J. J. and L. Q. Huang. 1991. Sorptive reversibility of atrazine and metolachlor residues in field soil samples. *J. Environ. Qual.* 20:222–228.

13. Pignatello, J. J. 1989. Sorption dynamics of organic compounds in soils and sediments. In *Reactions and Movement of Organic Chemicals in Soils*, B. C. Sawhney and K. Brown (ed), Soil Sci. Soc. Amer. Publ. 22.

14. Karickhoff, S. W. 1980. Sorption kinetics of hydrophobic pollutants in natural sediments. p. 193–205. In *Contaminants and Sediments*, R. A. Baker (ed.), Vol. 2. Ann Arbor Sci., Ann Arbor, MI.

15. Karickhoff, S. W. and K. R. Morris. 1985. Sorption dynamics of hydrophobic pollutants in sediments suspensions. *Environ. Toxicol. Chem.* 7:246–252.

16. Selim, H. H., J. M. Davidson, and R. S. Mansell. 1976. Evaluation of a two-site adsorption-desorption model for describing solute transport in soils. p. 44–448. In Proc. Summer Computer Simulation Conf. 12–14 July. Washington, DC.

17. Wu, S. and P. M. Gschwend. 1986. Sorption kinetics of hydrophobic organic compounds to natural sediments and soils. *Environ. Sci. Technol.* 20:717–725.

18. Steinberg, S. M., J. J. Pignatello, and B. L. Sawheny. 1987. Persistence of 1,2 dibromoethane in soils: entrapment in intraparticle micropores. *Environ. Sci. Technol.* 21:1201–1208.

19. Hance, R. J. 1967. The speed of attainment of sorption equilibria in soil systems involving herbicides. *Weed Res.* 7:29–36.

20. Wauchope, R. D. and R. S. Meyers. 1985. Adsorption-desorption kinetics of Atrazine and Linuron in freshwater-sediment aqueous slurries. *J. Environ. Qual.* 14:132–136.

21. Kahn, S. U. 1982. Studies on bound ^{14}C-prometryn residues in soils and plants. *Chemosphere* 11:771–795.

22. Spillner, C. J., J. R. DeDaun, and J. J. Menn. 1979. Degradation of fenitrothion in forest soils and effects on forest microbes. *J. Agric. Food Chem.* 27:1054–1060.

23. Subba-rao, R. V. and M. Alexander. 1982. Effect of sorption on mineralization of low concentrations of aromatic compounds in lake water samples. *Appl. Environ. Micro.* 44:659–688.

24. Turco, R. F. and A. E. Konopka. 1990. Response of microbial populations to carbofuran in soils enhanced for its degradation. American Chemical Society. In *Enhanced Biodegradation of Pesticides in the Environment*, K. D. Racke and J. R. Coats (ed).

25. Gerst, L. Z. and B. Yaron. 1983. Behavior of bromacil and napropamide in soils. I. Adsorption and desorption. *Soil Sci. Soc. Am. J.* 47:474–478.

26. Bromilow, R. A., R. J. Baker, M. A. H. Freeman, and K. Gorg. 1980. The degradation of Aldicarb and Oxamyl in soils. *Pest Sci.* 11:371–378.

27. Walker, A. 1978. Simulation of the persistence of eight soil-applied herbicides. *Weed Res.* 18:305–313.

28. Allen, R. and A. Walker. 1987. The influence of soil properties on the rate of degradation of metamitron, metazachlor and metribuzin. *Pest. Sci.* 18:95–111.

29. Weber, J. B. and H. D. Coble. 1968. Microbial decomposition of Diquat, adsorbed on Montmorillonite and Kaolinite clays. *J. Agr. Food Chem.* 16:475–477.

30. Ogram, A. V., R. E. Jessup, Litiou and P. S. C. Rao. Effects of sorption on biological degradation rates of (2,4-Dichlorophenoxy) acetic acid in soils. *App. Envir. Micro.* 49:582–587.

31. Senesi, N. and C. Testini. 1982. Physicochemical investigations of interaction mechanisms between *s*-triazines herbicides and soil humic acids. *Geoderma.* 28:129–146.

32. Mean, J. C. and R. Wiyayaratne. 1982. Role of natural colloids in the transport of hydrophobic pollutants. *Science* 215:968–970.

33. Carter, C. W. and I. H. Suffet. 1982. Binding of DDT to dissolved humic materials. *Env. Sci. Tech.* 16:735–740.

34. Jury, W. A., H. Elabd, and M. Resketo. 1986. Field studies of napropamide movement through unsaturated soil. *Water Resour. Res.* 22:749–755.

35. Mac Rae, I. C. 1986. Formation and degradation of soil bound ^{14}C-Fenitrothion residues in two agricultural soils. *Soil Biol. Biochem.* 18:221–225.

36. Rao, P. S. C. and J. M. Davidson. 1980. Estimation of pesticide retention and transformation parameters required in non-point source pollution models. p. 23–67. In *Environmental Impact of Nonpoint Source Pollution*, M. R. Overcash and J. M. Davidson (ed.), Ann Arbor Sci. Publ. Ann Arbor, MI.

37. Kahn, S. U. and H. A. Hamilton. 1980. Extractable and bound (non-extractable) residues of prometryn and its metabolites in an organic soil. *J. Agric. Food Chem.* 28:126–132.

38. Hurle, K. and A. Walker. 1980. Persistence and its prediction. In Interactions Between Herbicides and the Soil, R. J. Hance (ed), Chapter 4, pp. 83–122. Academic Press, London.

39. Guth, J. A. 1980. The study of transformations. In Interactions Between Herbicides and the Soil, R. J. Hance (ed.), Chapter 5, pp. 123–157. Academic Press, London.

40. Curl, E. A. and B. Truelove. 1986. The Rhizosphere (Advanced Series in Agricultural Sciences 15). 288 p. Springer-Verlag, Berlin.

41. Alexander, M. 1981. Biodegradation of chemicals of environmental concern. *Science* 211:132–136.

42. Pettygrove, D. R. and D. V. Naylor. 1983. Metribuzin degradation kinetics in organically amended soil. *Weed Sci.* 33:267–270.

43. McCormick, L. L. and A. E. Hiltbold. 1966. Microbiological decomposition of atrazine and diuron in soil. *Weeds* 14:77–82.

44. Burge, W. D. 1969. Populations of dalapon-decomposing bacteria in soil as influenced by additions of dalapon or other carbon sources. *Appl. Microbiol.* 17:545–550.

45. Hsu, T. S. and R. Bartha. 1979. Accelerated mineralization of two organophosphate insecticides in the rhizosphere. *Appl. Enviro. Micro.* 37:36–41.

46. Seibert, K., F. Fuhr, and W. Mittelst. 1982. Experiments on the influence of roots and soils on 2,4-D degradation. *Landwirt F.* 35:5–13.

47. Cheng, H. H., F. Fuhr, H. J. Jarczyk, and W. Mittelstaedt. 1978. Degradation of Methabenzthiazuron in the soil. *Agric. Food. Chem.* 26:595–599.

48. Rovira, A. D., R. C. Foster, and J. K. Martin. 1979. Note on terminology: Origin, nature and nomenclature of the organic materials in the rhizosphere. In *The Soil Root Interface,* J. L. Harley and R. S. Russell (eds), Academic Press, London, New York, pp. 1–4.

49. Martin, J. K. 1977. Effects of soil moisture on the release of organic carbon from wheat roots. *Soil Biol. Biochem.* 9:303–304.

50. Mench, M., J. L. Morel, and A. Guckert. 1987. Metal binding properties of high molecular weight soluble exudates from maize (*Zea mays* L.) roots. *Biol. Fertil. Soils* 3:165–169.

51. Allan, D. L., M. P. Russelle, and C. J. P. Gourley. 1989. The fate of symbiotically fixed nitrogen: A direct assessment in alfalfa. American Society of Agronomy 81st Annual Meetings.

52. Hurle, K. 1978. Einfluss des Strohverbrennens auf Akivität, Sorption und Abbau von Herbiziden in Boden. *Med. Fac. Landbouww. Rijksoniv. Gent.* 43:1097–1107.

53. Moyer, J. R., R. J. Haner, and C. E. McKone. 1972. The effect of absorbants on the rate of degradation of herbicides incubated in soil. *Soil Biol. Biochem.* 4:307–311.

54. Loux, M. M., R. Libel, and F. Slife. 1989. Bioactivity and persistence of Imazaquin. *Weed Sci.* 37:259–267.

55. van Genuchten, M. Th. and W. A. Jury. 1987. Progress in unsaturated flow and transport modeling. *Rev. Geophysics* 25:135–140.

56. Thomas, G. W. and R. E. Phillips. 1979. Consequences of water movement in macropores. *J. Environmental Quality* 8:149–152.

57. Bowman, R. S. and R. C. Rice. 1986. Accelerated herbicide leaching resulting from preferential flow phenomena and its implications for ground water contamination. In Proc. Conf. on Southwestern Groundwater Issues, Phoenix, AZ. 20–22 Oct. Natl. Water-Well Assoc., Dublin, OH.

58. Butters, G. L., W. A. Jury, and F. F. Ernst. 1989. Field scale transport of bromide in an unsaturated soil. I. Experimental methodology and results. *Water Resour. Res.* 25:1575–1582.

59. van Genuchten, M. Th. and P. J. Wierenga. 1976. Mass transfer studies in sorbing porous media. I. Analytic solutions. *Soil Sci. Soc. Amer. J.* 40:473–480.

60. van Genuchten, M. Th. and R. J. Wagenet. 1989. Two-site/two-region models for pesticide transport and degradation: 1. Theoretical. *Soil Sci. Soc. Amer. J.* 53:1303–1310.

61. Gamerdinger, A. P., R. J. Wagenet, and M. Th. van Genuchten. 1990. Application of two-site/two-region models for studying simultaneous nonequilibrium transport and degradation of pesticides. *Soil Sci. Soc. Amer. J.* 54:957–963.

62. Jury, W. A. 1982. Simulation of solute transport using a transfer function model. *Water Resour. Res.* 18:363–368.
63. Amoozegar-Fard, A., D. R. Nielsen, and A. W. Warrick. 1982. Soil solute concentration distributions for spatially varying pore water velocities and apparent diffusion coefficients. *Soil Sci. Soc. Am. J.* 46:3–8.
64. Jury, W. A. and M. Ghodrati. 1989. Overview of organic chemical environmental fate and transport modeling approaches. pp. 271–304. In *Reactions and movement of organic chemicals in soils*. Soil Sci. Soc. Am. Special Pub. No. 22.
65. Clendening, L. D. 1985. Evaluation of mobility screening parameters under field conditions. M.S. Thesis, Univ. of Calif., Riverside.
66. Ghodrati, M. 1989. The influence of water application method, surface preparation and pesticide formulation method on the preferential flow of pesticide through soil. Ph.D. Dissertation., Univ. of Calif., Riverside.
67. Gish, T. J., W. Zhuang, C. S. Helling, and P. C. Kearney. 1986. Chemical transport under no-till field conditions. *Geoderma* 38:251–259.
68. Boddy, P. L. and J. L. Baker. 1990. Conservation tillage effects on nitrate and atrazine leaching. Amer. Soc. Agric. Engin. Paper No. 90–2503.
69. Fermanich, K. J. and T. C. Daniel. 1991. Pesticide mobility and persistence in microlysimeter soil columns from a tilled and no-tilled plot. *J. Environ. Qual.* 20:195–202.
70. Richter, G. and W. A. Jury. 1986. A microlysimeter field study of solute transport through a structured sandy loam soil. *Soil Sci. Soc. Am. J.* 50:863–868.
71. Shipitalo, M. J., W. M. Edwards, W. A. Dick, and L. B. Owens. 1990. Initial storm effects on macropore transport of chemicals. *Soil Sci. Soc. Am. J.* 54:1530–1536.
72. Seyfried, M. S. and P. S. C. Rao. 1987. Solute transport in undisturbed columns of an aggregated tropical soil: Preferential flow effects. *Soil Sci. Soc. Am. J.* 51:1434–1444.
73. White, R. E., J. S. Dyson, Z. Gerstl, and B. Yaron. 1986. Leaching of herbicides through undisturbed cores of a structured clay soil. *Soil Sci. Soc. Am. J.* 50:277–283.
74. Richard, T. L. and T. S. Steenhuis. 1988. Tile drain sampling of preferential flow on a field scale. *J. Contam. Hydrol.* 3:307–325.
75. Cushman, J. H. 1984. On unifying the concepts of scale, instrumentation, and stochastics in the development of multiphase transport theory. *Water Resour. Res.* 20:1668–1676.
76. Cushman, J. H. 1986. On measurement, scale, and scaling. *Water Resour. Res.* 22:129–134.
77. Logan, T. J., G. W. Randall, and D. R. Timmons. 1980. Nutrient content of tile drainage from cropland in the North Central Region. No. Central Reg. Res. Pub. No. 268. 16 pp.
78. Baker, J. L. and H. P. Johnson. 1981. Nitrate-nitrogen in tile drainage as affected by fertilization. *J. Environ. Qual.* 10:519–522.
79. Hallberg, G. R., J. L. Baker, and G. W. Randall. 1986. Utility of tileline effluent studies to evaluate the impact of agricultural practices on ground water. pp. 298–326. In *Proc. of Conference on Agricultural Impacts on Ground Water, Aug. 11–13, Omaha, NE.* National Water Well Assoc., Dublin, OH.
80. Kanwar, R. S., J. L. Baker, and D. G. Baker. 1987. Tillage and N-fertilizer management effects on ground water quality. Paper No. 87–2077, Amer. Soc. Agric. Engin., St. Joseph, MI.

81. Bottcher, A. B., E. J. Monke, and L. F. Huggins. 1981. Nutrient and sediment loadings from a subsurface drainage system. *Trans. Am. Soc. Agric. Engin.* 24:1221–1226.

82. van Ommen, H. C., M. Th. van Genuchten, W. H. van der Molen, R. Kijksma, and J. Hulshoj. 1989. Experimental and theoretical analysis of solute transport from a diffuse source of pollution. *J. Hydrol.* 105–225–251.

83. Everts, C. J., R. S. Kanwar, E. C. Alexander, Jr., and S. C. Alexander. 1989. Comparison of tracer mobilities under laboratory and field conditions. *J. Environ. Qual.* 18:491–498.

84. Kladivko, E. J., G. E. Van Scoyoc, E. J. Monke, K. M. Oates, and W. Pask. 1991. Pesticide and nutrient movement into subsurface tile drains on a silt loam soil in Indiana. *J. Environ. Qual.* 20:264–270.

85. Kladivko, E. J., G. E. Van Scoyoc, R. F. Turco, and E. J. Monke. 1990. Tracer and pesticide transport through a silt loam soil into subsurface tile drains. *Agron. Abs.* 82, 41.

86. Jury, W. A. 1975a. Solute travel-time estimates for tile-drained fields. I. Theory. *Soil Sci. Soc. Am. Proc.* 39:1020–1024.

87. Jury, W. A. 1975b. Solute travel-time estimates for tile-drained fields. II. Application to experimental studies. *Soil Sci. Soc. Am. Proc.* 39:1024–1028.

88. Utermann, J., E. J. Kladivko, and W. A. Jury. 1990. Evaluating pesticide migration in tile-drained soils with a transfer function model. *J. Environ. Qual.* 19:707–714.

CHAPTER 5

Remediation of Pesticide Contaminated Soil at Agrichemical Facilities

Thomas J. Bicki and Alan S. Felsot

Landfarming may be a cost-effective, environmentally sound technology for remediation of pesticide-contaminated soil at agrichemical facilities. Remediation of a facility involves two components: (1) cleanup of the contaminated site and (2) treatment of the contaminated soil. Detailed remedial action plans are needed to facilitate proper removal and application of pesticide-contaminated soils. Case studies from two facilities in Illinois reveal that accurate assessment of the extent of contamination at the site and accurate characterization and application of the excavated soil at the landfarmed site are critical, if landfarming is to be a viable, environmentally sound method of remediation. Additionally, prolonged persistence of some compounds and crop phytotoxicity of landfarmed soil may be potential problems.

There are an estimated 14,000 agrichemical facilities in the U.S. that store, sell, and/or mix and apply pesticides and fertilizers.[1] Inadequate containment of pesticide spills and improper disposal of pesticide rinsates and containers at some facilities have resulted in point source contamination of soil and ground water. Contamination of soil and ground water resulting from past and present management, handling, and disposal practices at agrichemical facilities currently ranks among the highest of national environmental concerns.[2]

Soil and ground water contamination at agrichemical facilities often involves multiple pesticides at significantly higher concentrations than those normally associated with non-point source contamination. For example, an evaluation of ten agrichemical facilities in Iowa revealed that six commonly used herbicides were found in pools of water or in the soils of loading and handling areas at concentrations ranging from 1000 to 270,000 ppm.[3]

0-87371-926-3/94/$0.00+$.50
© 1994 by CRC Press, Inc.

An investigation of 20 agrichemical facilities in Wisconsin revealed the presence of at least 17 pesticides in soil and ground water samples with concentrations in soil reported as high as 37,950 ppm.[4] Several pesticides were frequently detected in ground water at concentrations exceeding health advisories (HA) or maximum contaminant levels (MCLs). Soils from acute spill areas, burn piles, and pesticide-impregnated fertilizer loading areas contained the greatest concentrations. Soils from pesticide mixing/loading areas and drainage ways had the widest array of compounds.

In Illinois, 77% of the water samples from 56 randomly selected wells at agrichemical facilities contained detectable residues of at least one pesticide.[5] Altogether, 14 pesticides were detected in well water samples and mean concentrations of several pesticides exceeded 140 ppb. Back-siphoning, improper mixing and loading procedures, lack of rinsate collection, and improper waste disposal practices were cited as possible causes of ground water contamination. Pesticide-contaminated soil at agrichemical facilities was noted as a potential "pesticide reservoir" that could lead to continual contamination of ground water in spite of any changes in operational procedures at the agrichemical facilities.

ROLE OF LANDFARMING AS A REMEDIATION TECHNOLOGY

A number of remedial technologies are currently available or are being developed to facilitate cleanup of pesticide-contaminated soil and ground water. These include physical, chemical, and biological treatment of excavated soil or excavation and disposal in an approved hazardous waste landfill. Physical remediation technologies include incineration, air stripping, mechanical aeration/extraction, activated carbon adsorption, soil flushing/washing, ion exchange, and membrane separation.[1] Remedial chemical treatments include neutralization, reduction-oxidation, photolysis, and dechlorination. Remedial action involving biodegradation includes biostimulation (the enrichment of waste with nutrients or organic substrates to stimulate degradation by indigenous microorganisms), bioaugmentation (the inoculation of wastes with microorganisms adapted for the degradation of a particular waste), or a combination of both.[6]

The design of a remedial action plan for a specific facility should begin with a thorough review of the available remedial action techniques. Each technique must be assessed in relation to the characteristics of the contaminated site and the contaminants present.

Unfortunately, many of the proposed remediation technologies are experimental and still in development. Furthermore, many agrichemical facilities are classified as small businesses and costs for cleanup and treatment are prohibitive. Landfarming (or land treatment) has been studied as an alternative technology because it is cost effective and does not require the development of new technology.[7-9]

Landfarming entails the application of pesticide-contaminated soil to agricultural land at agronomic application rates. It is a relatively old technology that has been used for treatment of municipal waste water and petroleum refinery sludges.[10] The objective of landfarming is the emplacement of pesticide wastes within the plow layer of the soil where dilution lowers the pesticide concentration sufficiently to facilitate both chemical and aerobic microbial degradation.

Landfarming is currently restricted to soils contaminated with pesticides currently registered for agronomic use. Most states that permit landfarming require contaminated soils containing Resource Conservation Recovery Act (RICRA) wastes or non-registered pesticides to be disposed of in hazardous waste landfills or by incineration.

Illinois was one of the first states to facilitate legislation to permit landfarming of pesticide-contaminated soil as a remediation practice. Legislation enacted in 1990 and effective until July 1, 1992 amended the Illinois Pesticide Act to allow the Illinois Department of Agriculture (IDOA) to permit an owner or operator of an agrichemical facility to remediate a site by land-applying pesticide-contaminated soil to agricultural land at agronomic rates. The Illinois Environmental Protection Agency (IEPA) has instituted an internal procedure to prescribe cleanup objectives and protocols for remedial action at agrichemical facilities.[11] The IDOA has developed operational control practices or protocols that are required when an applicant requests written authorization to land apply pesticide-contaminated soils to agricultural land.

SETTING CLEANUP OBJECTIVES

Conceptually, remediation of a facility can be divided into cleanup of the contaminated site and treatment of the contaminated soil. Before remediation of agrichemical facilities can be initiated on a large scale, however, a number of fundamental issues must be addressed. For example, how does one determine the need for remediation, or, if cleanup is initiated, how does one determine the end point of the remediation?

Most agricultural states, with the exception of Wisconsin, have not implemented formal written standards that address remediation of agrichemical facilities; sites are reviewed on a case-by-case basis. Contamination problems at agrichemical facilities may be so complex and site-specific as to essentially defy the practical application of traditional standards-setting procedures.[11] Some states use field application rate equivalents as cleanup objectives for soil.[4,12] In Illinois, health-based numbers such as the MCLs or HAs are proposed as cleanup objectives.[11] In some situations, a treatability factor of 3 to 5 times the MCL or HA may be employed, slightly raising the permissible residue levels in soil.

In a review of Wisconsin agrichemical facility investigations, Habecker[4] noted that contaminant trigger levels mandating remedial action at facilities have not been uniformly applied, and three approaches for setting remedial action trigger levels have been used. An early method involved the setting of a single trigger level for all pesticides. Trigger levels ranged from 1 to 10 ppb and were based on pesticide concentrations expected to be found in surface soils of agricultural fields following normal rates of pesticide application. Recently, Wisconsin adopted a new approach in which two trigger levels are set for a facility. One level is set for surface soil samples and a second, lower level is established for all subsurface soil samples. A third method for setting remedial action trigger levels incorporates the dual trigger level method plus the establishment of a trigger level based on the cumulative concentrations of all pesticides found at all depths at any one location.

DEVELOPMENT OF A REMEDIAL ACTION STRATEGY

A remedial action strategy must be developed before landfarming of pesticide-contaminated soil from agrichemical facilities is undertaken. Problems can arise when contamination extends beyond facility property lines. Fixing responsibility for remedial action becomes difficult when multiple properties are contaminated and when origin of the contamination is uncertain. The design of a remedial action plan for a specific facility should begin with an accurate assessment of the extent of soil and ground water contamination at a site. Protocols must be established to facilitate site assessments at both the remediation site and landfarming site. The areal extent of soil contamination, toxicity and occupational exposure to contaminants, sitting and installation of monitoring wells, volume of soil that must be excavated, excavation procedures, and physical and chemical properties of excavated soil must be components of the remedial action plan. The identities, concentrations, phytotoxicity, and degradation rates of pesticide residues are additional components needed in the remedial action plan. If replacement soil is needed to fill the excavated site, a source of fill must be identified, and pesticide residue analysis should be performed prior to backfilling to preclude reintroduction of contaminated soil to the site. Selection of backfill materials should also take into account the potential for recontamination of the site should another spill occur.

Assessment of a site selected to receive pesticide-contaminated soil should include components such as location of the site and distance from the source facility, property ownership, contractual agreement with landowner(s), vulnerability assessment of surface water and ground water resources, potential for offsite effects (i.e., leaching, runoff, volatilization, water and wind erosion potential, etc.). Additional components include identification of municipal, community or private wells, proposed travel route, demarcation of application area and field borders, siting locations of monitoring wells if needed, identification of physiographic or geomorphic landscape position (i.e., upland, terrace, floodplain, etc.), soil property characterization (i.e., organic matter content, clay content, infiltration rate, hydraulic conductivity, slope, etc.), antecedent pesticide residue levels in the soil, past and present tillage and cropping history and subsequent agronomic cropping practices, and timing of application.

Pretreatment of the excavated soil may be necessary prior to field application, and protocols should be included in the remedial action plan. Drying, removal of coarse fragments (e.g., boulders, cobbles, debris, etc.), grinding, blending, and mixing may be required and should be specified in the remedial action plan. Pesticide residue analysis of the pretreated soil must be performed to accurately characterize the kinds and concentrations of pesticide residues present so that application rates can be determined. Decontamination of the pretreatment site and equipment may also be needed.

Different approaches have been used to determine the appropriate field application rate of pesticide-contaminated soil. One approach is to calculate the application rate of contaminated soil based on the maximum concentration of each pesticide residue detected in soil borings. Another approach is to determine the maximum concentration of pesticide residues in excavated and mixed soil. Mixing of the excavated soil results in a lower maximum pesticide concentration in the soil. Lower maximum pesticide residue concentrations in mixed soil facilitates higher application rates of contaminated soil and smaller application site requirements.

Pretreatment at the application site may also be necessary. Marking of the actual application area and borders, seeding and establishment of vegetative borders, soil fertility and pH adjustments, selection of primary and secondary tillage treatments, field leveling, and residue management are additional considerations.

An incorporation depth for the contaminated soil should be specified and selection and calibration of application and incorporation equipment must be made. Pesticide residues levels at the application site should be documented immediately after application, and a schedule of follow-up pesticide residue analysis should be developed. Timing of contaminated soil application should take into account the optimal time for degradation of the various constituents, the slower degradation rates of pesticide residues in contaminated soil, phytotoxic effects on subsequent crops, and registration label restrictions on subsequent crops.

Finally, a closure of activity plan should document any problems encountered during the project, dates of the various activities, and analytical results of fill soil and field soil background levels, and field closure soil sampling plans.

On-going research in Illinois continues to examine and evaluate the feasibility of land application of pesticide-contaminated soil to agricultural land.[7-9] Case studies from remedial actions at two facilities in Illinois are presented to illustrate the scope and importance of proper assessment and characterization of a contaminated site and soil, pretreatment and application considerations, and an assessment of degradation rates of several herbicides following landfarming.

CASE STUDIES OF AGRICHEMICAL FACILITY REMEDIATION

Case Study A — Galesville, Illinois

Improper discharge of pesticide rinsate at an agrichemical facility in the town of Galesville in Piatt County, Illinois led to contamination of approximately 802 m^3 of soil. The IEPA ordered remediation of the site when soil, along a railroad right-of-way adjoining the facility, was found to be contaminated with unacceptably high levels of herbicides (e.g., 24,000 ppm alachlor in the top 10 cm). Analyses of composite samples collected from the upper 60 cm of soil near the former rinsate discharge point indicated average concentrations of 110 ppm alachlor, 62 ppm atrazine, and 20 ppm metolachlor. To avoid disposal of the contaminated soil at a hazardous waste landfill, we collaborated with the IEPA and the management of the facility to landfarm the contaminated soil.[7] The landfarming site was located immediately adjacent to the facility and was owned by the facility.

During the planning of the landfarming experiment, four criteria for successful remediation were developed: (1) no significant differences, after one growing season, between herbicide residues in soil from waste-treated plots and freshly sprayed plots; (2) no significant toxicity, as measured by phytotoxicity assays, in the field or greenhouse and by comparison to yields from the untreated check plots; (3) no significant residues in grain; and (4) no contamination of shallow ground water above MCLs suggested by the U.S. EPA.

In situ contaminated soil at the site was separated into three zones to facilitate excavation of soil with varying pesticide residue concentrations (Figure 1). Soil from each zone was excavated separately with a bulldozer and dredge scoop, mixed, and placed into designated waste soil piles. Waste piles were sampled twice before application to determine the pesticide constituents and their concentrations and to calculate appropriate loading rates of contaminated soil on adjacent cropland. The contaminated soil contained a mixture of herbicides, including alachlor, metolachlor, atrazine, and trifluralin.

Persistence of herbicide residues in landfarmed, contaminated soil was compared to persistence of herbicides that were freshly sprayed with amounts calculated to yield concentrations in soils similar to those in herbicide-contaminated soil. Rates of application were determined on the basis of alachlor concentration because it was the most prevalent contaminant. Application rates for the treatments applied reflect kg of active ingredient applied per hectare. Varying amounts of freshly sprayed herbicides and herbicide-contaminated soil were applied to corn and soybean plots in 1986. Treatments were arranged in a randomized complete block design with three replications (Figure 2). Treatments were designated by the following codes:

1. CHECK: untreated soil;

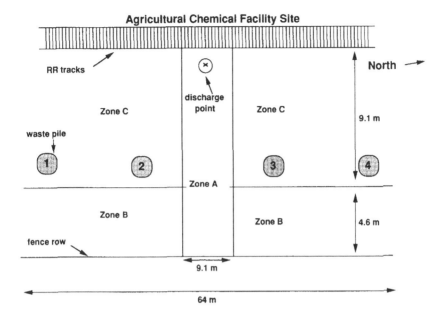

FIGURE 1 Schematic of pesticide-contaminated site in Galesville, IL. Soil was excavated and stored in piles numbered 1 to 4. Soil from pile 2 was used in the landfarming experiments.

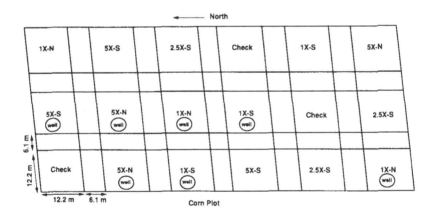

FIGURE 2 Experimental design of landfarming experiment (randomized, complete block with three replications per treatment). See text for meaning of treatment codes. Wells were placed in the designated plots. The soybean plot had only one well, which was placed in the CHECK treatment.

2. 1X-N: herbicide spray mixture applied at 3.4 kg/ha rate, with mixture consisting of alachlor, metolachlor, atrazine, and trifluralin in proportion to the concentrations found in contaminated soil;

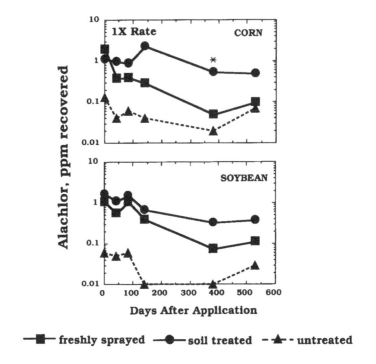

FIGURE 3 Alachlor residues in corn and soybean plots following a 1× equivalent alachlor application rate (EAAR).

3. 5X-N: herbicide spray mixture applied at 16.8 kg/ha rate (i.e., 5 times the recommended alachlor rate);
4. 1X-S: contaminated soil applied at equivalent alachlor application rate (EAAR) of 3.4 kg/ha;
5. 2.5X-S: contaminated soil applied at EAAR of 8.4 kg/ha;
6. 5X-S: contaminated soil applied at EAAR of 16.8 kg/ha.

Dissipation of herbicides in soil derived from freshly sprayed products and contaminated soil were monitored for 528 days following application. Sampling protocol, analytical methods, and herbicide residues detected in soil from individual sampling periods have been previously described in detail.[7,8] During the first 140 days of sampling, high variability sometimes precluded the detection of significant differences between corresponding rates of application of freshly sprayed herbicide and contaminated soil-applied herbicide. Despite the high interplot variability in residue recovery, concentrations of alachlor and metolachlor were generally larger in soil-treated plots than in freshly sprayed plots (Figures 3 to 6).

Significantly more alachlor and metolachlor was recovered from soil (corn plots only) in the 1X-S treatment than from soil in the 1X-N treatment, 380 days after application (Figures 3 and 4). However, by 528 days, herbicide

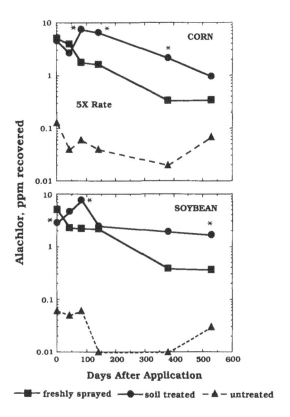

FIGURE 4 Alachlor residues in corn and soybean plots following a 5× EAAR.

recovery in soil was not significantly different in 1X-S and 1X-N treatments. Significantly more alachlor and metolachlor was recovered in soil (corn plots only) in the 5X-S treatment than from soil in the 5X-N treatment, 380 days after application (Figure 5 and 6). By 528 days, recovery of alachlor (soybeans only) and metolachlor in soil was significantly higher in 5X-S than in 5X-N treatments.

In field bioassays, corn was not adversely affected by herbicide residues, but soybean injury was noted in 5X treatments, probably as a result of the excessive atrazine concentrations.[7] Although 95% of the soybeans were injured in 5X-N plots, injury in 5X-S plots averaged only 20%. Because residue recovery of atrazine in soybeans plots did not significantly differ by treatment, the phytotoxicity assays suggested differences in bioavailability between the residues in soil-treated and freshly sprayed plots.

In phytotoxicity assays conducted in the greenhouse, soybeans and corn were injured when grown in contaminated soil collected in 1986 but not in soil collected in 1988.[8,9] Grain yields were not affected by application of contaminated soil, and bioaccumulation of parent herbicide residues in grain did not occur.

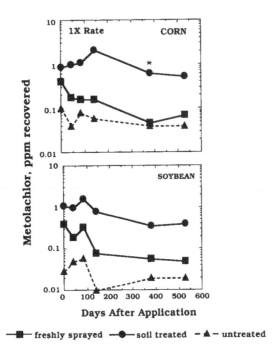

FIGURE 5 Metolachlor residues in corn and soybean plots following a 1× EARR.

Although herbicide residues were detected in shallow ground water,[8,9] no significant differences in concentrations were found among treatments throughout 140 days of monitoring. Initial well water samples, collected from the field prior to landfarming of contaminated soil, showed that ground water had already been contaminated as a result either of past farming practices or migration from the adjacent agrichemical facility; therefore, effects of landfarming on ground water quality at the application site could not be evaluated.

Not all contaminated soil excavated from the facility was utilized in the landfarming treatments. Soil remaining in waste piles was sampled 6 months, 1 year, and 2 years after excavation to determine whether residue concentrations had declined as a result of mixing, aeration, and aging of the soil.[8]

Minimal degradation of herbicides in the excess waste-pile soil was noted 140 days after excavation. Alachlor did dissipate by about 60%, within 1 year after excavation, but the residues stabilized at 20 to 25 ppm (oven-dry soil basis) the following year. Metolachlor was a comparatively minor constituent in the waste pile following excavation, but its concentration was similar to the concentration of alachlor after 2 years. Atrazine degraded in waste-pile soil to levels of 2 to 3 ppm. Trifluralin did not dissipate in the waste-pile soil, remaining at about 3 ppm.

After 2 years of monitoring herbicide persistence and biological activity at

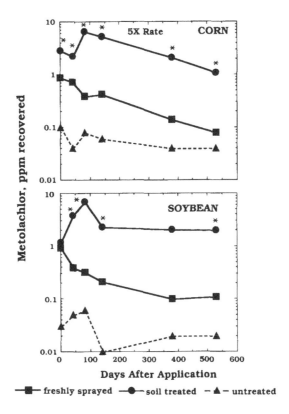

FIGURE 6 Metolachlor residues in corn and soybean plots following a 5× EARR.

the landfarmed site, it was concluded that prolonged persistence of herbicide residues in the soil and crop phytotoxicity are potential problems with land application of herbicide contaminated soil.

Case Study B — Lexington, Illinois

Our experiences with a warehouse fire and the aftermath of firefighting procedures effectively illustrate problems of site assessment and soil remediation, which are necessary to prevent ground water contamination. In April 1990, a pesticide warehouse caught fire within the town of Lexington in McLean County, Illinois. Water used to fight the fire became visibly contaminated by herbicides leaking from fire-damaged containers. The contaminated water began to flow away from the warehouse structure and entered a drainage tile that ran under a nearby field and discharged to surface water. To prevent further runoff of water, the tile inlet was sealed, and earthen berms were placed around the warehouse; the contaminated water was contained between the warehouse and adjacent structures, an automobile dealership on the east and a grain elevator on the west. The color of the water indicated contamination with

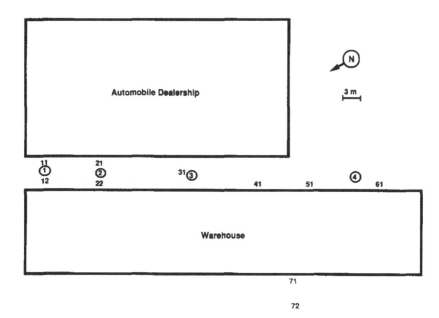

FIGURE 7 Schematic diagram of sampling points for soil cores at the site of the pesticide warehouse fire in Lexington, IL. Circled numbers represent samples collected with a hand-operated bucket auger (data in Table 1); other numbers represent samples collected by split spoon sampler (data in Table 2).

one or more herbicides. Eventually the water, contained by the berm and buildings, infiltrated into the soil. The top 5 cm of soil in certain areas around the warehouse had a greenish-yellow color which suggested to us the presence of trifluralin. The IEPA ordered a cleanup of contaminated soil located between the buildings. Soil sampling was carried out to assist the IEPA in determining the depth of herbicide leaching. Landfarming of the contaminated soil was advised, and degradation of the most prevalent contaminant, trifluralin, was monitored.

Soil around the warehouse was sampled two times in late April, approximately 2 weeks after the fire. During the first sampling period, only soil on the east side of the warehouse was sampled because this was believed to be the site of accumulation of the most contaminated water runoff (Figure 7). A 5-cm-diameter hand-operated bucket auger was used to collect samples. Samples were collected from four locations in 15-cm increments to a depth of 90 cm. The upper 1 to 2 cm of soil in the auger was discarded to avoid contamination from shallower depths. Four days later, more extensive sampling was conducted on both sides of the warehouse (Figure 7) using a hydraulic split spoon sampler in a hollow stem-auger. Samples were taken to a depth of 360 cm in 30-cm increments. Obvious contamination from overlying soil layers was usually noticed in the upper portion of the core within the split spoon, and efforts were made to exclude this material during subsequent analytical procedures.

Table 1. Recoveries of Herbicides in Soil Cores Collected
23 April 1990 at the Site of a Pesticide
Warehouse Fire in Lexington, Illinois[a]

Sample site	Depth (cm)	Parts per billion (ppb)		
		Trifluralin	Atrazine	Alachlor
1	0–15	25,077	13,019	1,340
1	15–30	507	396	0[b]
1	45–60	933	412	0
1	75–90	363	284	72
2	0–15	1,753	3,184	1,507
2	15–30	39	257	0
2	45–60	434	219	0
2	75–90	360	170	0
3	0–15	717,323	610,650	0
3	15–30	99,598	48,628	0
3	45–60	6,361	4,748	103
3	60–75	22,107	24,430	92
3	75–90	13,141	5,021	680
4	0–15	2,896	579	98
4	15–30	5,180	760	184
4	30–45	12,276	1,299	280
4	45–60	4,193	728	96
4	60–75	1,317	425	0

[a] Soils collected by hand auger.
[b] 0 ppm indicates that no GLC response was detected at the appropriate retention time for alachlor. Limits of determination under the conditions of extraction (50-g soil with a final volume of 10 ml) and GLC were 40, 20, and 80 ppb for trifluralin, atrazine, and alachlor, respectively.

Upon return to the laboratory, soils were frozen at –20°C for 1 week or less before extraction. After thawing, soils were ground and mixed by hand. Then 50-g aliquots were slurried with water and extracted twice by stirring with 90 ml of ethyl acetate.[7,8] After concentration of the solvent, extracts were analyzed for trifluralin, atrazine, alachlor, metolachlor, and cyanazine using a 90-cm column of Apiezon plus DEGS.[13] Because cyanazine and metolachlor were not completely resolved on this column, the concentration of these herbicides was not quantitated.

Initial sampling of soil on the east side of the warehouse by hand augering indicated significant contamination by trifluralin and atrazine to a depth of 90 cm; alachlor concentrations were comparatively minor (Table 1). Herbicide concentrations declined with distance from site 3. Because initial observations indicated significant contamination at 90 cm (i.e., concentrations were greater than agronomic rates of use), further sampling was required to delineate the depth of contamination.

Based on analyses of soil collected with the split spoon sampler, the highest concentrations of trifluralin and atrazine were distributed in the upper 90 cm; below this level herbicide concentrations dropped significantly (Table 2). Trifluralin was the most predominant contaminant and was most concentrated at sites 31, 41, and 51 (Figure 7). Using an alternative gas chromatographic

Table 2. Recoveries of Herbicides in Selected Soil Cores
          Collected 27 April 1990 at the Site of a Pesticide
          Warehouse Fire in Lexington, Illinois[a]

Sample site	Depth (cm)	Parts per billion (ppb)		
		Trifluralin	Atrazine	Alachlor
12	0–30	22,707	5,98	594
12	30–60	3,344	961	66
12	60–90	1,511	517	197
12	90–120	515	129	23
12	120–150	828	610	145
12	150–180	36	51	0[b]
12	180–210	62	34	13
12	210–240	100	25	0
22	0–30	24,937	48,050	1,217
22	30–60	1,148	1,010	52
22	60–90	1,197	791	0
22	90–120	1,846	351	0
22	120–150	989	933	71
22	150–180	68	94	184
22	180–210	0	0	0
22	210–240	16	18	0
31	0–30	22,805	1,379	0
31	30–60	22,641	3,062	0
31	60–90	78,277	4,867	523
31	90–120	5,050	805	0
31	120–150	11,247	722	0
31	150–180	5,152	598	111
31	180–210	3,058	1,253	0
31	210–240	1,303	368	71
31	240–270	976	157	25
31	270–300	252	5	0
31	300–330	628	17	0
31	330–360	1,001	131	97
41	0–30	58,198	1,903	10,545
41	30–60	18,222	288	0
41	60–90	26,751	2,500	4,695
41	90–120	2,249	212	0
41	120–150	1,199	206	0
41	210–240	866	ND[c]	0
41	240–270	ND	ND	ND
41	270–300	3,979	ND	0
41	300–330	2,729	138	ND
41	330–360	1,244	30	0
51	0–30	38,698	11,620	2,338
51	30–60	105,508	2,579	3,732
51	60–90	19,763	459	0
51	90–120	111	58	0
51	120–150	225	74	0
51	150–180	610	92	0
51	180–210	114	8	0
51	210–240	160	11	0
51	240–270	2,239	77	0
51	270–300	415	13	0
61	0–30	267	622	242
61	30–60	173	132	0
61	60–90	53	35	23
61	90–120	190	72	0

Table 2. Recoveries of Herbicides in Selected Soil Cores
Collected 27 April 1990 at the Site of a Pesticide
Warehouse Fire in Lexington, Illinois[a] (continued)

Sample site	Depth (cm)	Parts per billion (ppb)		
		Trifluralin	Atrazine	Alachlor
61	120–150	260	17	0
61	150–180	ND	46	0
61	180–210	74	27	216
61	210–240	0	33	0
71	0–30	11,843	523	0
71	30–60	177	60	0
71	60–90	84	17	0
71	90–120	<LD[d]	51	0
71	120–150	64	22	<LD
71	150–180	90	47	0
71	180–210	140	49	<LD
71	210–240	37	9	0

a Soils were collected using hollow stem auger and split spoon
   sampler.
b 0 ppb indicates that no GLC response was detected at the
   appropriate retention time for trifluralin, atrazine, or alachlor.
c ND indicates identification of peak was not determined as a
   result of interfering peaks that made identification or quantitation
   invalid.
d Limit of Determination; 8, 4, and 20 ppb for trifluralin, atrazine,
   and alachlor, respectively, under the conditions of extraction
   (50-g soil, 2 ml final volume) and gas-liquid chromatography.

technique (60-cm column of Ultrabond with temperature programming),[14] we were able to confirm that metolachlor was also present in high concentrations in the top 90 cm at site 41 (up to 6700 ppb in the top 30 cm, data not shown).

Despite the rapid decrease in herbicide concentrations below 90 cm, analyses of deeper soil cores revealed significant movement to a depth of 360 cm at sites 31, 41, and 51. Efforts were made to exclude contamination of lower soil cores by soil from the top of the profile, but occasionally contamination was unavoidable. The core depths that had comparatively higher probabilities of contamination were those ending at 90, 150, 210, 270, and 330 cm (i.e., the top 30 cm of the 60-cm split spoon corer). With some deep samples, the suspected contaminating soil was separated from the core and extracted separately. The results showed that soil contamination from the top of the profile could contribute significantly to the concentration of herbicide in the deep core (perhaps up to a tenfold increase in concentration, data not shown). Contamination of cores ending at depths of 60, 120, 180, 240, 300, and 360 cm (i.e., the bottom 30 cm of the 60-cm core) was less likely to have occurred, and concentration of herbicides in these cores more accurately reflected the leaching pattern of the herbicides.

As observed in the soils collected by bucket auger, collection with the split spoon showed that contamination dropped off rapidly with distance from sites 31 and 51 (Figure 7, Table 2). Sampling points to the north (site 11) and south (site 61) were significantly less contaminated below 30 cm, which suggested a gradient of decreasing subsurface contamination away from the central part of the warehouse. This portion of the warehouse was most heavily damaged by the fire. On the west side of the warehouse, concentrations of trifluralin and atrazine were generally lower than corresponding samples on the east side (compare samples 71 with 41 and 51 in Table 2). This side of the warehouse was covered with crushed limestone; furthermore, there was more room for firefighting water to spread out and enter a nearby tile drain before the drain was closed off.

Visual examination of the deep cores indicated that a water table was absent within the depth of sampling, and the subsoil materials did not correspond to known shallow aquifer materials. Based on the absence of a water table at the depths of sampling and the rapid drop in herbicide concentration below 90 cm, the IEPA permitted excavation of soil on the east side of the warehouse to a depth of 90 to 120 cm.

In early May 1990, approximately 300 tons of contaminated soil was transported to a farm about 5 miles away and stockpiled until August 1990. The waste piles, 60 to 150 cm high, encompassed an area of approximately 97.5 m by 34.7 m. To assess the average concentration of contaminants within the waste piles, the area was divided into four quadrants. In July 1990, four 5-cm diameter cores were randomly collected from the top 10 cm of the piles in each quadrant and composited. Analyses of the soil indicated that the concentration of trifluralin ranged from 84 ppm to 4050 ppm. The latter concentration was found in quadrant 3; a core had been taken from soil that was noticeably green in color, which suggested to us the presence of large amounts of trifluralin. Disregarding the outlier from quadrant 3, the average concentration of trifluralin was 134 ± 47 ppm. Other contaminants in the soil included atrazine, alachlor, metolachlor, and cyanazine at 10 to 100-fold lower concentrations.

Quadrant one was repeatedly sampled in August 1990 while a backhoe was loading soil into a manure spreader during the landfarming treatment. Six samples each (10 cm × 4 cm diameter) were taken from the top 8 cm, the 90 to 120 cm depth, and the 150 cm depth of the soil piles. Trifluralin concentration in individual soil cores ranged from 3 ppm to 1003 ppm with a mean of 158 ± 247 ppm. Although the means from the two sampling dates were similar, variation was greater when individual cores were analyzed from within one quadrant. The experience with sampling and analysis of pesticides in waste soil piles showed the necessity of repeated sampling on two different days. Thorough assessment of the mean concentration of contaminants is a prerequisite to calculating the load of soil and acreage needed to achieve the mandated agronomic rate of application for the most limiting pesticide constituent (i.e., the pesticide at the highest concentration, or alternatively the pesticide most toxic to the intended crop).

In late August 1990 soil was spread over an approximately 80-acre field of "set-aside" land using two applications with a manure spreader. Soil in waste

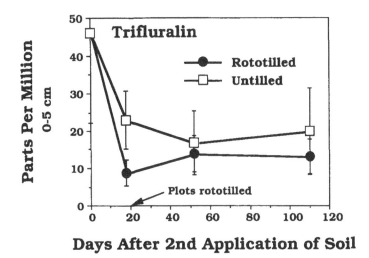

FIGURE 8. Recovery of trifluralin following application of contaminated soil to an uncropped field. The soil was applied in August 1990 and residues were monitored until December 1990. Plots were either rototilled or left untilled (four replicates of each treatment). Vertical lines represent standard deviations.

piles was applied without previous mixing. Application of the contaminated soil was under the direction of the warehouse manager and the exact soil loading rate was not controlled by us. Based on herbicide concentrations in stockpiled soil, our calculations indicated that a single application of the waste soil should not have exceeded the recommended agronomic application rate for trifluralin. After the second application, four samples were collected along a 60-m transect across the field to determine the initial concentration of trifluralin in the top 5 cm of soil.

To monitor the dissipation of trifluralin, eight plots were set up in the middle of the field during September. The plots were 3 m × 3 m; every other plot was rototilled to a depth of approximately 5 cm. The remaining plots were left untilled. Periodically random cores from each plot were collected and bulked together for laboratory analysis of trifluralin. Because trifluralin has a very low water solubility (0.3 ppm) and moderately high vapor pressure (1.1×10^{-4} mmHg),[15] it volatilizes quickly from soil; therefore, the experimental design tested the hypothesis that lack of soil incorporation by tillage would favor volatilization of trifluralin residues in the landfarmed soil.

The mean concentration of trifluralin in soil cores (5 cm deep) collected along the transect after the second pass by the manure spreader was approximately 45 ppm (Figure 8). At a recommended maximum application rate of 2.2 kg AI/ha, the concentration of trifluralin should be approximately 3.5 ppm in the top 5 cm. Thus, the rate of application exceeded the guidelines of the IDOA. The variance from the guidelines probably resulted from both lack of mixing of stockpiled soil prior to application and failure to calibrate the manure

spreader and uniformly apply the contaminated soil. Within about 30 days after
the plots were rototilled, however, the concentration of trifluralin had declined
to 10 to 15 ppm (Figure 8). Except on the day of rototilling, recovery of
trifluralin was not affected by tillage. Furthermore, as the soil temperature
declined throughout the fall, trifluralin dissipation slowed and seemed inhib-
ited. At this point, plans were made to continue sampling the plots during the
subsequent growing season.

The lack of further dissipation of trifluralin as soil temperatures cooled
illustrated the importance of timing to the success of a landfarming treatment.
Landfarming of pesticide-contaminated soils should start as early as possible
in the spring to allow exposure to maximum microbial activity. At this time,
tillage does not seem to be a factor in affecting the dissipation of comparatively
volatile compounds. However, disking of the contaminated soil into the field
should promote better contact with microbiologically active soil.

CONCLUSIONS

Landfarming may be a cost-effective, environmentally sound technology
for remediation of pesticide contamination at agrichemical facilities.
Remediation of a facility involves two components: (1) cleanup of the contaminated site and (2) treatment of the contaminated soil. Development of detailed
remedial action plans are needed to facilitate proper removal, application, and
degradation of pesticide residues.

Potential problems requiring further research include the prolonged persistence of some compounds and crop phytotoxicity. The two case studies presented illustrate that degradation of pesticide residues in contaminated soil may
be a function of the length of time residues have been present in the soil and
their concentration. Pesticides appear to degrade at a slower rate in soil that has
been contaminated for relatively long periods of time, (e.g., as in the case of
the Galesville site), while degradation of high concentrations in recently contaminated soil (e.g., as in the case of the Lexington site) appears to be slower
than the degradation expected from lower concentrations characteristic of
agronomic rates. Timing of landfarming may be a critical parameter to overcome reductions in microbial activity as soil temperature cools following the
growing season.

LITERATURE CITED

1. Norwood, V. M. and Randolph, M. E. *A Literature Review of Biological Treatment and Bioremediation Technologies Which May Be Applicable at Fertilizer/
   Agrichemical Dealer Sites*, National Fertilizer and Environmental Research Center, Muscles Shoals, AL, 1990; pp 7–15.

2. Anastos, G.; Corbin, M. H.; and Coia, M. F. *Toxic and Hazardous Waste* 1987, 19, 163–171.
3. Hallberg, G. R. In *Proceedings Agricultural Impacts on Ground Water — A Conference*, National Water Well Association, Dublin, OH, 1986; pp 1–66.
4. Habecker, M. A. *Environmental Contamination At Wisconsin Pesticide Mixing/ Loading Facilities: Case Study, Investigation, and Remedial Action Evaluation.* Wisconsin Department of Agriculture, Trade and Consumer Protection, Agricultural Resource Management Division, Madison, WI, 1989; pp 1–52.
5. Long, T. In *Proceedings of the 16th Annual ENR Conference*, Illinois Department of Energy and Natural Resources, Springfield, IL, 1988; pp 143–149.
6. Grubbs, J.; Piotrowski, M. R.; and Wilson, S. B. *Hazmat World* 1991, 4, 44–55.
7. Felsot, A.; Liebl, R.; and Bicki, T. J. *Feasibility of Land Application of Soils Contaminated With Pesticide Waste as a Remediation Practice*, HWRIC RR 021, Illinois Department of Energy and Natural Resources, Springfield, IL, 1988; 55 pp.
8. Felsot, A.; Dzantor, E. K.; Case, L.; and Liebl, R. *Assessment of Problems Associated With Landfilling or Land Application of Pesticide Waste And Feasibility of Cleanup By Microbial Degradation*, HWRIC RR 053, Illinois Department of Energy and Natural Resources, Springfield, IL, 1990; 68 pp.
9. Felsot, A. S. and Dzantor, E. K. In *Pesticides In The Next Decade: The Challenges Ahead, Proceedings of a National Research Conference*, Weigmann, D., Ed.; Virginia Water Resources Research Center, Blacksburg, VA, 1991; pp 532–551.
10. Soil Science Society of America. *Utilization, Treatment, and Disposal of Waste on Land*, American Society of Agronomy, Madison, WI, 1986; 318 pp.
11. Taylor, A. G. Discussion Paper: A Perspective Regarding the Determination of Cleanup Objectives and Protocols at Contaminated Agrichemical Retail Sites in Illinois. Illinois Environmental Protection Agency, Springfield, IL, 1990.
12. Frank, J. F. In *1991 Illinois Fertilizer Conference Proceedings*, Hoeft, R. G., Ed.; Cooperative Extension Service, University of Illinois, Urbana, IL, 1991; pp 41–44.
13. Felsot, A. S. and Dzantor, E. K. In *Enhanced Biodegradation of Pesticides in the Environment*, Racke, K. D. and Coats, J. R., Eds; American Chemical Society Symposium Series No. 426, American Chemical Society, Washington, D.C., 1990; pp 192–213.
14. Roseboom, D.; Felsot, A.; Hill, T.; and Rodsater, J. *Stream Yields From Agricultural Chemicals and Feedlot Runoff From an Illinois Watershed.* ILENR/RE-WR-90/11, Illinois Department Energy and Natural Resources, Springfield, IL, 1990; 133 pp.
15. Weed Science Society of America, *Herbicide Handbook* (sixth edition), Champaign, IL, 1989; pp 252–253.

CHAPTER 6

Modeling the Movement of a Rapidly Degrading Solute, Methomyl, in Dynamic Soil-Water Systems

Kathryn C. Dowling, Ronald G. Costella, and Ann T. Lemley

The movement of the carbamate insecticide methomyl, *S*-methyl *N*-(methyl-carbamoyloxy)thioacetimidate, under steady-state flow was investigated in a small laboratory soil column. This dynamic flow system permitted the simultaneous study of sorption and degradation interactions that occur during transport. Miscible displacement experiments along with mass balance calculations, moment analysis, and batch incubation studies furnished independent information on insecticide degradation and sorption. Methomyl decay under miscible displacement was rapid (half-life values are on the order of several hours to a day), first-order, and faster for higher organic carbon soils. Higher organic carbon soils retain methomyl to a somewhat greater degree, similar to the effect observed with aldicarb, but muted in comparison with that seen in previous work on the insecticide ethoprop and the triazine herbicides. High degradation rates and some non-equilibrium sorption characterize methomyl soil interactions. Solutions to the convection-dispersion equation were used for quantitative evaluation of transport processes. The solution that includes terms for first-order degradation and equilibrium sorption was most applicable but was unable to completely describe methomyl's complex soil interactions.

As of 1988, ground water, an important source of drinking water in the U.S., was found to be contaminated in 26 states with 46 pesticides linked to normal agricultural practices.[1] The U.S. EPA's 1990 National Survey of Pesticides[2] estimates that about 10% (10,000 nationwide) of community water system wells and 4% (450,000 nationwide) of rural domestic wells contain detectable levels of at least 1 pesticide. The 1988 EPA Interim Report includes findings of methomyl in New York state ground water.[1] Methomyl, *S*-methyl *N*-(methylcarbamoyloxy) thioacetimidate, with a chemical structure very similar to that of the problematic aldicarb, is highly water soluble and therefore likely to leach, and could pose a significant threat to ground water.

Existing methods for miscible displacement experiments, as modified in our laboratory, have been used in previous studies to evaluate the extent of aldicarb,[3] ethprop,[4] and triazine[5] leaching through soils. The current study of methomyl confirms the usefulness of this methodology as a rapid screening tool and further tests the applicability of mathematical models to the description of insecticide movement in soil. This project's objectives include (1) the determination of sorption in different soils under static (batch) conditions, (2) the examination of the effect of varying experimental conditions on transport and degradation under dynamic flow conditions, and (3) the evaluation of the equilibrium model's ability to describe column displacement results.

MATERIALS AND METHODS

Methomyl ($\geq$99% pure) was obtained from USEPA, Pesticides and Industrial Chemicals Repository, Research Triangle Park, NC and Chem Service, West Chester, PA. Potassium chloride (Mallinckrodt, Paris, KY) served as the miscible displacement experimental tracer. The solvent acetonitrile (HPLC grade, Fisher Scientific, Fair Lawn, NJ) was employed in chromatographic insecticide analyses. All aqueous insecticide solutions prepared for column or batch studies contained 0.005 M $CaSO_4$ (reagent grade, Mallinckrodt, Paris, KY) in distilled, reverse osmosis (DRO) water to mimic soil water electrolytic properties.

Methomyl batch study samples were filtered with Supelco or Whatman nylon syringe-tip filters (0.45-μm pores, 3 or 25 mm diameters) to remove suspended soil particles. Control studies were performed to confirm that methomyl was not adsorbed by the nylon filters. HPLC analysis was performed on a Hewlett Packard 1090A HPLC equipped with a Supelco LC-8-DB column (3-μm particle size, 4.6 mm i.d. $\times$ 15 cm) at 50°C and a Supelguard LC-8-DB guard column. The diode-array detector was set at 235 nm, and the mobile phase (25/75% acetonitrile/DRO water) flow rate was 1 ml min^{-1} with an injection size of 1 ml.

Table 1 describes the three soils used: Riverhead (a loam/sandy loam), Rhinebeck (a silty clay loam), and Valois (a silt loam). The Rhinebeck and Valois soils were collected from Tompkins County, New York, in Fall 1985, and the Riverhead soil was collected from field 2E at the Long Island Horticultural Research Laboratory, Riverhead, New York, in July 1986. Samples were taken from the 5 to 30 cm depth, air-dried, and sieved to a size of $\leq$2 mm for batch and soil column studies.

Batch Studies

The batch method, comprised of a biphasic system of soil and water in a capped 50-ml centrifuge tube, was used to quantify insecticide sorption to a

Table 1. Soil Characteristics

Property	Riverhead	Rhinebeck	Valois
Bulk density (g/cm^3)	1.22	0.95	1.15
pH	4.4	6.7	5.9
Cation exch. cap. (meq/100 g)	11.5	27.0	15.0
Exch. acidity H$^+$ (meq/100 g)	12.0	9.0	11.0
Organic matter (%)	2.3	5.9	3.2
Organic carbon (%)	1.07	3.13	1.64
Texture			
Percent sand	51	12	30
Percent silt	36	52	55
Percent clay	12	36	15

particular soil. A soil/solution ratio of 1:1 was used. Preliminary experiments were conducted to determine at what time sorption was complete but degradation had not yet occurred. Based on these results, incubation studies were conducted for 23 and 24 h for methomyl in Riverhead soil, for 17 and 18 h in Valois soil, and for 4 and 8 h in Rhinebeck soil.

For all batch studies, 10 ± 0.04 g soil was weighed into 50-ml capped centrifuge tubes to which 10 ml of insecticide solution of either the concentration range 10, 4, 2, 1, and 0.5 mg/l or the range 20, 13, 10, 4, 2, 1, and 0.5 mg/l in 0.005 M CaSO$_4$ was added. After 30 min of agitation followed by centrifugation (10 min at 3200 rpm), the aqueous layer was removed and analyzed for insecticide concentration. The soil layer was extracted using DRO water; the extracts were centrifuged and analyzed.

Miscible Displacement Experiments

Miscible displacement techniques, as refined by Zhong et al.,[6] involve the application of a pulse of insecticide solution to a soil column and the periodic collection of aqueous effluent fractions that are analyzed for insecticide concentration, yielding a breakthrough curve (BTC). Beckman Instruments glass columns (2.5 cm i.d. × 21.5 cm) with Teflon end fittings were packed with dry soil tamped to a density somewhat higher than the measured soil bulk density. Complete saturation of each column with the electrolyte solution (0.005 M CaSO$_4$) was accomplished with a Sage Instruments continuous delivery syringe pump model 220 at a flow rate of 0.26 cm^3 min^{-1}. The flow rate was then adjusted to the desired experimental velocity and allowed to equilibrate.

In initial experiments, electrolyte solution was cleared from the syringes, and a discrete pulse of methomyl solution was applied to the soil column. In later experiments (M4, M6–M8, M10, and M11), a second syringe pump was loaded with the insecticide solution, and the pulse was applied using a Hamilton HV3-3 three-port switching valve. Constant temperature was maintained for all experiments at 25 ± 2°C either in a temperature-controlled room or using a circulating water jacket surrounding the column. An Isco Retriever II or a

Buchler LC-100 fraction collector was used to sample column effluent at timed intervals.

Experimental conditions of soil type, flow rate, and pulse length were varied and are given in Table 2. Flow rates and pulse lengths were controlled with the continuous delivery syringe pump and were measured experimentally; these values were confirmed by computer analysis of chloride ion tracer effluent curves as described below. The chloride ion tracer (approximately 100 mg/l Cl⁻, analyzed on a Buchler-Cotlove Chloridometer equipped with silver electrodes) was run simultaneously with methomyl in all cases to establish the values of the hydrodynamic parameters: average pore water velocity, v (cm h⁻¹); apparent dispersion, D (cm² h⁻¹); and insecticide pulse duration, t_o (h).

Methomyl was completely eluted from the soil column in all experiments. Effluents rapidly reached zero insecticide concentration, and subsequent extractions of soil (carried out with DRO water) did not yield any recovery of methomyl. More detailed explanations of the above experimental methods are provided elsewhere.[4]

THEORETICAL

For water-soluble compounds such as methomyl, sorption in the static batch system can be described by the linear sorption isotherm $s = k_d c$, in which s (mg kg⁻¹) is the amount of the insecticide sorbed to the soil, c (mg l⁻¹) is the aqueous insecticide concentration at equilibrium, and k_d (l kg⁻¹) is the sorption coefficient for pesticide distribution between the soil and aqueous phases. Plots of s vs. c should be linear and intersect the origin.[7] Alternatively, the Freundlich relationship describes non-linear sorption isotherms:[7] $s = k_f c^{1/n}$, where n is an empirical constant.

Several mathematical models which provide analytical solutions to the convection-dispersion equation (CDE) have been developed to describe solute transport during miscible displacement and are reviewed elsewhere.[8-10] One of these solutions (previously used with aldicarb and its breakdown products)[3,6] was applied in this work to quantitatively describe experimental BTCs. The version of the CDE employed (Equation 1) describes flux-averaged concentrations, includes an equilibrium conceptualization of sorption, and incorporates first-order degradation. The model was used to evaluate the movement of both the tracers and the insecticide methomyl.

$$(1 + \rho k_d/\theta)\partial c/\partial t = D\partial^2 c/\partial z^2 - v\partial c/\partial z - \mu c \tag{1}$$

In Equation 1, D is the apparent dispersion coefficient (cm² h⁻¹) and v is the average pore water velocity (cm h⁻¹). The first-order degradation rate constant, the soil bulk density, and the volumetric water content are represented by μ (h⁻¹), ρ (g cm⁻³), and θ (cm³ cm⁻³), respectively. This model assumes that degradation occurs only in the

Table 2. Experimental Conditions and Tracer-Determined, Moment Analysis, and CXTFIT-Calculated Parameters for Methomyl Miscible Displacement Experiments

Experimental Conditions

	M1	M2	M3	M4
Soil	Ri	Ri	Ri	Ri
ρ (g cm^{-3})	1.4	1.5	1.4	1.4
θ (cm^3 cm^{-3})	0.42	0.37	0.45	0.42
c (mg l^{-1})	10.2	10.1	9.7	9.9
	M5	**M6**	**M7**	**M8**
Soil	Va	Va	Va	Va
ρ (g cm^{-3})	1.2	1.2	1.3	1.3
θ (cm^3 cm^{-3})	0.49	0.50	0.47	0.47
c (mg l^{-1})	8.9	10.4	9.6	9.8
	M9	**M10**	**M11**	**M12**
Soil	Rh	Rh	Rh	Rh
ρ (g cm^{-3})	1.1	1.1	1.1	1.1
θ (cm^3 cm^{-3})	0.50	0.51	0.51	0.52
c (mg l^{-1})	10.1	9.9	10.1	10.6

Tracer-Determined Parameters

	M1	M2	M3	M4
v (cm h^{-1})	8.9	7.0	1.2	0.9
D (cm^2 h^{-1})	12.3	2.4	3.2	3.0
t_o (h)	4.4	3.2	30.1	46.6
	M5	**M6**	**M7**	**M8**
v (cm h^{-1})	9.1	8.0	3.3	1.0
D (cm^2 h^{-1})	5.7	12.4	2.0	1.1
t_o (h)	3.8	6.3	12.3	48.2
	M9	**M10**	**M11**	**M12**
v (cm h^{-1})	7.1	5.7	4.4	0.8
D (cm^2 h^{-1})	14.0	7.3	13.2	2.9
t_o (h)	2.4	10.0	15.4	37.5

CXTFIT-Calculated and Moment Analysis Parameters

	M1	M2	M3	M4
CXTFIT k_d[a] (l kg^{-1})	0.055 (0.045,0.065)	0.027 (0.021,0.033)	<0 (−0.052,−0.017)	<0 (−0.23,0)
CXTFIT μ[a] (h^{-1})	0 (0,0)	0.017 (0.002,0.032)	0.028 (0.025,0.031)	0.079 (0.059,0.099)
Moment analysis k_d (l kg^{-1})	0.057	0.040	b	b
Mass Balance μ (h^{-1})	0	0	8.7×10^{-3}	0.026
	M5	**M6**	**M7**	**M8**
CXTFIT k_d (l kg^{-1})	0.11 (0.10,0.12)	0.22 (0.20,0.27)	0.052 (0.042,0.067)	c
CXTFIT μ (h^{-1})	0.041 (0.028,0.053)	0.11 (0.089,0.11)	0.15 (0.14,0.15)	c
Moment analysis k_d (l kg^{-1})	0.14	b	b	b
Mass balance μ (h^{-1})	5.6×10^{-3}	0.030	0.040	c
	M9	**M10**	**M11**	**M12**
CXTFIT k_d (l kg^{-1})	0.11 (0.097,0.12)	<0 (−0.27,−0.071)	c	c

Table 2. Experimental Conditions and Tracer-Determined, Moment Analysis, and CXTFIT-Calculated Parameters for Methomyl Miscible Displacement Experiments (continued)

CXTFIT μ (h^{-1})	0.25 (0.23,0.26)	0.40 (0.37,1.1)	c	c
Moment analysis k_d ($l\ kg^{-1}$)	b	b	b	b
Mass balance μ (h^{-1})	0.085	0.094	c	c

[a] Numbers in parentheses reflect the 95% CT for the above value.
[b] Moment analysis was only performed on experiments with no or virtually no degradation.
[c] These experiments yielded no BTCs so these parameter values were not estimated.

solution phase. The retardation factor, R, is related to linear insecticide sorption and is defined as $R = 1 + \rho k_d/\theta$.

The parameter optimization code CXTFIT is based on an analytical solution[11] to the CDE shown above (Equation 1). CXTFIT uses a curve-fitting method based on a minimization of the sums of squares of the residuals to estimate parameter values. When analyzing insecticide BTCs, the program requires initial estimates of D, v, and t_o; these are provided by CXTFIT analyses of chloride ion tracer BTCs based on assumptions of Cl$^-$ values for R and μ (see below). Thus, methomyl BTCs were analyzed with values for v, t_o, and D fixed. In this way, CXTFIT yielded estimates of R and μ for the insecticide.

Use of chloride ion as a non-interactive tracer was previously studied[4,5] by comparing it in each experimental soil with tritiated water (^3H_2O), which is assumed to lack preferential interactions with the soil (R = 1) and to be non-degrading ($\mu = 0$). Chloride ion, also non-degrading, was found to be slightly excluded compared to tritiated water; corrected chloride retardation values for each soil were determined and used in all subsequent tracer BTC analyses. Chloride retardation values are 0.94 and 0.92 for Riverhead soil at high (5 to 10 cm h^{-1}) and low (0 to 5 cm h^{-1}) velocities;[4] 0.84 and 0.81 for Rhinebeck soil at high and low velocities;[4] and 0.91 and 0.92 for Valois soil[12] at high and low velocities, respectively.

The moment analysis technique described by Valocchi[13] was used to estimate model-independent k_d values for experiments exhibiting no or virtually no degradation (M1, M2, and M5).

RESULTS AND DISCUSSION

Batch Studies

The sorption isotherms for methomyl in the experimental soils were linear; all R^2 values are 0.99 or greater (Table 3). The degree of sorption is as one would predict based on the methomyl K_{ow} value of 1.2[14] and soil organic

Table 3. Sorption Parameters for Methomyl Batch Systems

Soil	% OCC	Incubation time (h)	Concentration range (ppm)	k_d (cm^3/g)	R^2
Present Study					
Riverhead	1.07	23	0.5–20	0.15	0.99
Riverhead	1.07	24	0.5–10	0.30	1.00
Valois	1.64	17	0.5–20	0.17	1.00
Valois	1.64	18	0.5–10	0.33	1.00
Rhinebeck	3.13	4	0.5–20	0.39	1.00
Rhinebeck	3.13	8	0.5–10	0.43	0.99
Literature Values[a15]					
Naaldwijk	3.7	4	0.25, 1.0	0.46	—
Honseler	3.8	4	0.25, 1.0	0.43	—
Aalsmeer	9.7	4	0.25, 1.0	1.30	—

[a] The values given as % OCC are actually % OMC since these authors did not report % OCC values. Each of the two concentrations shown was run in duplicate so four points were generated for each curve.

carbon contents (OCC). Methomyl is highly water soluble and its sorption coefficients (k_d) are therefore low. In addition, the degree of methomyl sorption increases for higher OCC soils.

Experimental methomyl sorption coefficients agree with literature values from Leistra et al.[15] for various soils (Table 3). Ali et al.[16] conducted 4-h batch studies over a methomyl concentration range of 5 to 30 mg/l on a 0.4% and a 3.0% organic matter content (OMC) soil. Methomyl sorption to the 0.4% OMC soil was found to deviate from Freundlich's relationship; the sorption isotherm was non-linear even under Freundlich analysis. In the 3.0% OMC soil, the methomyl sorption isotherm conformed to the Freundlich relationship although an n value of 0.5 demonstrated the isotherm was non-linear. Thus, the literature indicates that in some soils methomyl sorption isotherms are not linear; however, they were linear in the present study.

Methomyl Miscible Displacement Studies

Methomyl BTCs in Riverhead soil at high velocities (M1 and M2; Figure 1) demonstrate rapid elution and a small amount of asymmetry, an indication that non-equilibrium sorption plays a minor role. When the BTC data were analyzed using a deterministic partial equilibrium/non-equilibrium two-site sorption solution incorporating first-order degradation[12] (parameter optimization code CXT4), results confirmed that equilibrium sorption prevailed. For most cases the CXT4 code did not converge or yield parameter estimates with confidence intervals that overlapped zero. In those cases for which CXT4 parameter estimates were non-zero, the majority (60 to 80% depending on the soil) of sorption sites were found to be at equilibrium. Thus, the equilibrium sorption solution (CXTFIT) was chosen to analyze the BTCs.

Although no decay was measured in the M1 and M2 high velocity experiments, it was detected in the Riverhead lower velocity experiments (M3 and M4; Figure 2), as evidenced by the greatly reduced areas under the BTCs.

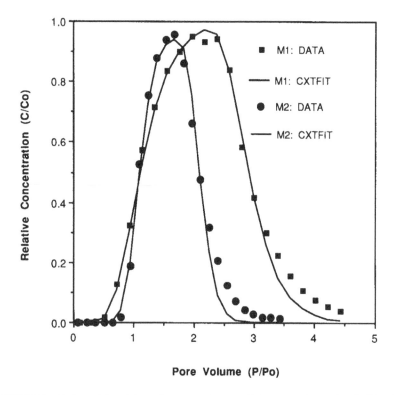

FIGURE 1 Methomyl breakthrough curves for two higher experimental velocities in Riverhead soil.

These BTCs, particularly curve M4 where the pulse was applied for 46.6 h, show asymmetric character. The results of our studies indicate that the CXTFIT and CXT4 algorithms do not reach satisfactory solutions for the asymmetric curves characteristic of experiments with pulse lengths greater than the insecticide half-life.

The high velocity experiments in Valois soil (M5 and M6) are shown in Figure 3. Unlike the high velocity experiments in Riverhead soil, degradation was measured for both of these experiments. The different areas under the BTCs attest to different amounts of overall degradation. Methomyl's rapid decay in Rhinebeck soil (M9 and M10; Figure 4) is also evidenced by decreased BTC areas. The M6 and M10 pulses were applied for longer periods than in experiments M5 and M9 (see Table 2), and there is considerable asymmetry in both BTCs. Three experiments at the lowest flow rates on both Rhinebeck (M11 and M12) and Valois soils (M8) yielded no detectable breakthrough curves at all; the applied methomyl was entirely degraded on-column over the course of the experiments.

For all experiments producing BTCs, the model-independent experimental mass balance results for μ, the degradation rate constant, are included in Table 2.

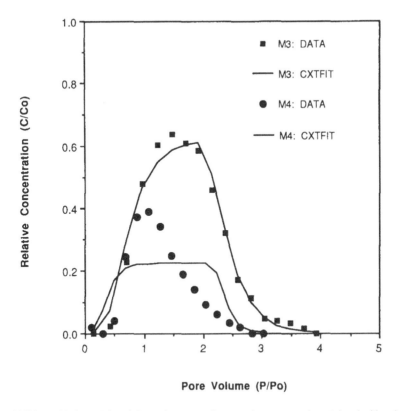

FIGURE 2 Methomyl breakthrough curves for two lower experimental velocities in Riverhead soil.

These results confirm that degradation occurs in those experiments with decreased areas under the BTCs. The mass balance rates can be compared for different soils. Methomyl decay in Riverhead soil could only be observed for low velocity experiments; M4 demonstrates a half-life of about 1 day. The decay detected in the Valois soil (M6 and M7) corresponds to a half-life of about 20 h; for the Rhinebeck high velocity experiments (M9 and M10), half-life values are about 8 h. These methomyl degradation rates, measured by miscible displacement, increase with increasing soil OCC and predict faster degradation than do the preliminary batch studies. No measurable degradation occurred during the period of the batch studies (4 to 24 h depending on the soil), a result that corresponds to the work of Aly et al.[17] who found half-life values of 8 days for methomyl in non-autoclaved 0.4% OMC soil.

CXTFIT parameter estimates for methomyl are presented in Table 2. As discussed above, CXTFIT does not model the data as well as expected. All methomyl BTCs were resolved by allowing the model to come to a solution for the retardation and degradation parameters. Ideally, the value for retardation is calculated from the batch sorption coefficient and used as a fixed input parameter

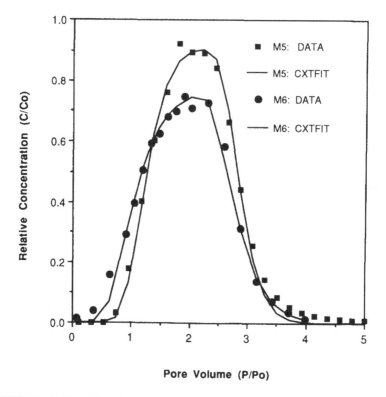

FIGURE 3 Methomyl breakthrough curves for two higher experimental velocities in Valois soil.

so CXTFIT can estimate the rate constant. The fits were poor, however, using this latter method (labeled as "CXTFIT R FIXED"; see Figure 5 for an example). The estimated BTC displays a distinct shift to the right using this approach. In general, fits were better if retardation was not fixed (the case labeled as "CXTFIT" in Figure 5).

Although allowing CXTFIT to estimate both R and μ provides reasonably good fits for experiments in which degradation rates are low (M1 to M3, M5, M6, and even M9), this approach fails to provide good fits for experiments with higher degradation (M4, M7, and M10). Poor fits of portions of the curves (i.e., the tails) of the former class of BTCs can probably be attributed to a small amount of non-equilibrium sorption. The latter class of experiments, with lower velocities and lengthy pulse applications, exhibits asymmetric BTCs that CXTFIT is unable to resolve effectively. It is important to note that even in experiments for which fits appear good, parameter estimates for degradation are consistently several times higher than model-independent rate constants derived from mass balance calculations, and estimates for sorption are several times lower than batch-derived values (Table 2). In extreme cases (M3, M4, and M10), parameter estimates for retardation are so low as to yield sorption

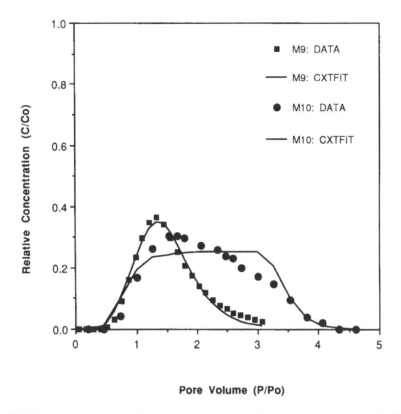

FIGURE 4 Methomyl breakthrough curves for two higher experimental velocities in Rhinebeck soil.

coefficients of less than zero. Moment analysis of BTC curves M1, M2, and M5 confirms CXTFIT estimates for the sorption coefficients (Table 2). Since batch methodology was designed specifically to preclude degradation (although not time-dependent sorption), batch results represent true sorption values uncomplicated by degradation effects. We hypothesize that degradation occurring during column studies affects BTC shapes such that sorption estimates appear lower than values measured in the batch system. The curve-fitting techniques employed here (the CXTFIT code accompanied by confirmatory moment analysis) do not adequately describe the combination of degradation and sorption occurring in the column system. In addition, these techniques are not able to confirm the batch sorption coefficient values measured.

In summary, CXTFIT does not appear able to describe adequately the interactions of methomyl with the experimental soils even in the case of experiments with no degradation. Deficient BTC fitting may be due to a combination of factors not properly addressed by the program, including high rates of degradation and the presence of small amounts (less than 40%) of non-equilibrium sites.

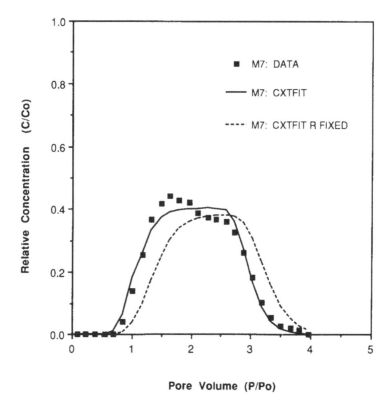

FIGURE 5 A methomyl breakthrough curve for a moderate experimental velocity in Valois soil showing the effect of two different curve-fitting strategies.

CONCLUSIONS

Batch studies demonstrate that methomyl sorption is higher for higher OCC soils. Methomyl also is retained to a somewhat greater degree in those miscible displacement studies involving higher OCC soils. The degree of increased retention with increased soil OCC is less than for other more hydrophobic pesticides. For all three soils studied, methomyl rapidly eluted from the column. Methomyl decay in soil column systems appears to increase with increasing soil OCC and is rapid (with half-life values ranging from several hours to 1 day), especially considering that these are laboratory and not field studies.

The parameter optimization code CXTFIT does not adequately describe the combination of high degradation and non-equilibrium sorption that characterizes methomyl interactions with soil. Sorption coefficients are underestimated and degradation rates are overestimated. Model-independent estimates of sorption coefficients confirm CXTFIT results, whereas model-independent estimates of degradation fall between batch and CXTFIT column results. These differences are especially evident for experiments in which a high proportion of methomyl applied to the soil column degrades over the course of the experiment. Since the partial equilibrium/non-equilibrium two-site sorption solution (CXT4) also does not effectively describe methomyl behavior in miscible displacement experiments, more advanced models must be developed to evaluate pesticides that exhibit high degradation and have small amounts of non-equilibrium sorption.

In conclusion, the relative threat of methomyl to ground water can be assessed using the data obtained in this study. Methomyl rapidly decomposes, especially in higher OCC soils, and moves rapidly through lower OCC soils. Methomyl may be more problematic in lower organic content soils, where it is retained to a lesser degree. Thus, ground water contamination by methomyl is more likely in areas with low organic content soils and high precipitation, which would increase methomyl movement through soil and decrease methomyl-soil interactions.

LIST OF ABBREVIATIONS

BTC	breakthrough curve
c	aqueous insecticide concentration (mg l^{-1})
CDE	convection-dispersion equation
CXTFIT	parameter optimization code: one-site solution
CXT4	parameter optimization code: two-site solution
D	apparent dispersion coefficient (cm^2 h^{-1})
HPLC	high performance liquid chromatography
k_d	linear sorption coefficient (l kg^{-1})
k_f	Freundlich sorption coefficient
K_{ow}	octanol-water partition coefficient
n	Freundlich degree of linearity
OCC	soil organic carbon content
OMC	soil organic matter content
P	column pore volume (cm^3)
R	retardation factor (dimensionless)
s	sorbed insecticide concentration (mg kg^{-1})
t_o	insecticide pulse duration (h)
U.S. EPA	United States Environmental Protection Agency
v	average pore water velocity (cm h^{-1})
μ	degradation rate constant (h^{-1})
ρ	bulk density (soil) (g cm^{-3})
θ	volumetric water content (soil) (cm^3 cm^{-3})

ACKNOWLEDGMENTS

Dr. Amy P. Gamerdinger supervised Natalie Fredd who conducted two of the column experiments, and she and Dr. David M. Crohn made valuable suggestions about this work. Research funding was provided by the USDA Regional Research Fund, the Northeast National Pesticide Impact Assessment Program, the New York State College of Human Ecology at Cornell, and the United States Army.

LITERATURE CITED

1. Williams, W. M.; Holden, P. W.; Parsons, D. W.; and Lorber, M. N. *Pesticides in Ground Water Data Base: 1988 Interim Report,* U.S. EPA, December 1988.
2. *National Survey of Pesticides in Drinking Water Wells Phase I Report.* U.S. EPA Offices of Water and of Pesticides and Toxic Substances, November 1990, 570/9-90-015.
3. Lemley, A. T.; Wagenet, R. J.; and Zhong, W. Z. *J. Environ. Qual.* 1988, 17, pp 408–414.
4. Dowling, K. C. *The Transport and Degradation of Ethoprop and Methomyl in Soil-Water Systems,* M.S. Thesis, Cornell University, 1990.
5. Gamerdinger, A. P.; Lemley, A. T.; and Wagenet, R. J. *J. Environ. Qual.* 1991, 20, pp 815–822.
6. Zhong, W. Z.; Lemley, A. T.; and Wagenet, R. J. In *Evaluation of Pesticides in Ground Water;* Garner, W. Y., Honeycutt, R. C., and Nigg, H. N., Eds.; ACS Symposium Series 315; American Chemical Society: Washington, D.C., 1986; pp 61–77.
7. Cheng, H. H. and Koskinen, W. C. In *Evaluation of Pesticides in Ground Water;* Garner, W. Y., Honeycutt, R. C., and Nigg, H. N., Eds.; ACS Symposium Series 315; American Chemical Society: Washington, DC, 1986; pp 2–13.
8. Wagenet, R. J. and Rao, P. S. C. *Weed Sci.* 1985, 33 (Suppl. 2), pp 25–32.
9. Wagenet, R. J. In *Evaluation of Pesticides in Ground Water;* Garner, W. Y., Honeycutt, R. C., and Nigg, H. N., Eds.; ACS Symposium Series 315; American Chemical Society: Washington, D.C., 1986; pp 330–341.
10. Addiscott, T. M. and Wagenet, R. J. *J. Soil Sci.* 1985, 36, pp 411–424.
11. Parker, J. C. and van Genuchten, M. Th. *Determining Transport Parameters from Laboratory and Field Tracer Experiments;* Virginia Agricultural Experiment Station, 1984, Bulletin 84-3.
12. Gamerdinger, A. P.; Wagenet, R. J.; and van Genuchten, M. Th. *Soil Sci. Soc. Am. J.* 1990, 54, pp 957–963.
13. Valocchi, A. L. Validity of the local equilibrium assumption for modeling sorbing solute transport through homogeneous soils, *Water Resour. Res.* 1985, 21, pp 808–820.
14. Francis, B. M. and Metcalf, R. L. *Laboratory Model Ecosystem Evaluations of Twenty-six New Pesticides;* University of Illinois at Urbana-Champaign, Institute for Environmental Studies, 1981, Research Report No. 9, p 80.

15. Leistra, M.; Dekker, A.; and Van der Burg, A. M. M. *Water Air Soil Pollut.* 1984, 23, pp 155–167.
16. Ali, M. I.; Bakry, N.; Kishk, F.; El-Sebae, A. H.; and Abo-El-Amayaem, M. In *Proceedings of the Fourth Conference of Pest Control;* National Research Centre: Cairo, 1978; pp 555–563.
17. Aly, M. I.; Bakry, N.; El-Seabe, A. H.; and Elamayem, M. A. *Alex. J. Agric. Res.* 1979, 27, pp 689–697.

CHAPTER 7

Modeling the Degradation and Movement of Agricultural Chemicals in Ground Water

Russell L. Jones

During the past decade, considerable progress has been made in modeling the movement and degradation of agricultural chemicals in the environment. Usually simulations of agricultural residues in ground water must also include unsaturated zone simulations to predict the magnitude and timing of residue movement into ground water. Therefore, ground water modeling often involves the use of a linked system of unsaturated and saturated zone models. One of the most important considerations in ground water modeling is choosing the model to match the desired objectives. The use of complex ground water models is quite appropriate for interpreting the results of comprehensive field experiments. However, simple models are often appropriate for assessment simulations (not site specific) needed for regulatory decisions or developing management practices.

The discovery of agricultural chemical residues in drinking water supplies beginning in 1979[1-3] resulted in the initiation of research programs by industry, regulatory, and university scientists. These included laboratory experiments, field research and monitoring studies, and modeling research, including the development of computer models for describing the movement and degradation of agricultural chemicals in both the unsaturated and saturated zones. During the past 10 years, considerable progress has been made in modeling the movement of mobile agricultural chemicals.[4] Most of this progress has been in the development and validation of unsaturated zone models; however, saturated zone models have also been developed and used to assist in interpretation of field research and to develop management practices for protecting drinking water supplies.[5-8]

The occurrence of agricultural chemicals in drinking water wells is usually the result of the combined effects of two different types of movement. First,

0-87371-926-3/94/$0.00+$.50

117

under certain circumstances agricultural chemicals can move downwards through the soil profile and into shallow ground water underneath treated fields. Second, agricultural chemicals can move with ground water to nearby wells. Modeling the latter process is the focus of this paper.

BEHAVIOR OF AGRICULTURAL CHEMICALS IN THE SATURATED ZONE

Modeling the behavior of non-volatile agricultural chemicals in the unsaturated and saturated zones is conceptually similar. Three factors affect the movement of a non-volatile agricultural chemical in both soil and ground water. One of these factors is the degradation rate which is usually dependent on soil or aquifer properties (such as particle size, pH, organic matter, and microbial concentrations) and temperature. Another important factor is the movement of water. In the unsaturated zone, water movement is affected by the amount of rainfall, the amount and type of irrigation, evapotranspiration losses, and soil hydraulic properties (such as field capacity and wilting point). However, water movement in the saturated zone is quite site specific and dependent primarily on the geological characteristics of the site. The other important factor is the movement of the agricultural chemical relative to the movement of water. In the unsaturated zone, chemical movement is retarded by the sorption to soil particles. Since sorption to organic matter is the primary mechanism, this sorption in usually expressed using the organic matter content of a soil and a measured value for the organic carbon/water partition coefficient. Since little organic matter is normally found in the saturated zone, movement of agricultural chemicals in the saturated zone is usually essentially the same as the ground water movement.

Although unsaturated and saturated zone models are conceptually similar, in practice such models are usually quite different. The main differences include the way residues are introduced into the simulation and the differences in the calculation of the water flow. As a result of these differences, comprehensive saturated zone models are usually more complicated and require more input data and computer time to perform a simulation.

In an unsaturated zone simulation, agricultural chemicals are introduced into the soil by applications made at one or more specified times. In contrast, agricultural chemical residues enter into the saturated zone over a period of time usually in response to specific rainfall events. The amount and timing of this residue movement into ground water is usually unknown and also varies spatially within a field due to differences in soil types and water table depths. One common approach to the problem of estimating residue inputs into the saturated zone is the use of unsaturated zone models. One or more unsaturated zone simulations may be performed to estimate the amount and timing of residue movement to the water table.

Flow of water in the unsaturated zone can normally be described one dimensionally (exceptions include simulations where localized areas of saturation may result in lateral flow of water). In saturated zone modeling, two or three dimensions may be needed to describe temporal and spatially varying patterns of water movement in site-specific simulations.

MODEL SELECTION

Perhaps the most important step in saturated zone modeling is defining the overall objectives of the simulation, including the needed confidence interval around the simulation results. After the simulation goals are clearly defined, the modeler can then plan a suitable approach, deciding on which assumptions and simplifications should be made (considering also the applicable environmental conditions). Potential sources of error in saturated zone model predictions include the theory and assumptions that comprise a model, the numerical solution of model equations, and selection of input parameters. The modeler controls these errors by the choice of an appropriate model for the specified application and by correct specification of input parameters. Therefore, the modeler should be aware of the assumptions inherent in a model. When considering the use of a model for a certain application, the appropriateness of these assumptions should be examined. The appropriateness of extracting parameters from work conducted at other locations or with other compounds should also be reviewed.

The conclusions drawn from a set of modeling simulations is usually more dependent on the skill of the modeler, rather than the exact choice of models used in the simulations. The modeler must be able to develop a set of simulations that properly address the objectives and then have the computational skills needed to correctly perform these simulations. The demands for these two types of skills vary from problem to problem. When the objective is to validate a complex model using comprehensive field data, problem formulation skills of the modeler are relatively unimportant while the ability to perform a complex simulation is critical. However, if the objective is to develop an appropriate management practice for a given region (perhaps requiring only relatively simple model calculations), then problem formulation skills are critical while computational skills may be less important.

The most complex modeling simulations are usually site-specific simulations, where some field data are available. For such site-specific simulations (often performed for model validation or calibration), the ground water flow rate and direction usually varies throughout the site, often as a function of time. Such situations are best modeled using a two- or three-dimensional model with numerical solution techniques. However, when performing general assessments needed for regulatory purposes (such as estimating distances agricultural chemicals may potentially move from treated fields), only a range of typical

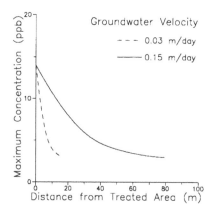

FIGURE 1 Predicted maximum aldoxycarb concentrations in ground water during a 17-year simulation period for a field near Harrellsville, NC. The calculations assume a 1.5-m water table, a saturated zone half-life of 6 months, an aquifer porosity of 0.3, a longitudinal dispersivity of 10 m, and that aldoxycarb residues in the saturated zone are concentrated in a 1.5-m vertical band. (Adapted from Reference 15.)

ground water velocities in the region may be available. In these simulations, probably the most appropriate model would be a simple one-dimensional model, using an analytical or simple numerical solution technique.

EXISTING MODELS

Although many different models have been developed for simulating movement of chemicals in ground water, most of this ground water research has focused on chemical movement from point sources such as landfills and leaching storage tanks. Some of these general models (for example, References 9 to 11) can be used to simulate non-point situations such as agricultural chemicals moving through the soil profile. To date, linked modeling systems developed or modified by the modeler have been used to conduct most of the ground water modeling research on agricultural chemicals involving movement in both the unsaturated and saturated zones. However, the EPA model RUSTIC[12,13] issued in 1989 does contain a saturated zone module which can be linked to other modules simulating movement of agricultural chemicals in the unsaturated zone (an example simulation is presented in Reference 14).

EXAMPLE APPLICATIONS

As discussed previously, simulations of the degradation and movement of agricultural chemicals in ground water can vary significantly in complexity, depending on the objectives and available data. The following examples from the author's experience attempt to convey this diversity.

Soil Specific

The objective of this simulation[15] was to estimate potential ground water concentrations of aldoxycarb following applications to North Carolina tobacco on a Norfolk loamy fine sand (the soil type at a test site near Harrellsville, NC). Figure 1 shows the maximum concentration predicted underneath and downgradient of the treatment area during a 17-year simulation. These predictions were made by coupling the unsaturated zone model PRZM[16,17] with a one-dimensional finite element model, the core of which is a simple solute transport code.[18]

Site Specific

These simulations[19] were performed to help estimate the degradation rate of aldicarb residues in ground water during an unsaturated and saturated zone research study conducted near Lake Hamilton, FL. Simulations were performed with a two-dimensional version of the simple finite element model used for the previous example, which had also been adapted to account for the spatial variation in water table depth underneath the treated area. Ground water elevation measurements made during the study were used to determine average ground water gradients for the 3.5-year simulation period. Field and laboratory measurements were used to characterize the behavior of aldicarb residues in the soil. Relatively good agreement was obtained between field measurements and model simulations.

Regulatory

Using the half-life estimate obtained in the previous example, modeling simulations were performed[19] to estimate potential movement of aldicarb residues downgradient of treated citrus groves located on the Florida ridge. Simulations were performed with the previously mentioned simple one-dimensional saturated zone model linked to PRZM for a range of degradation rates and water table depths (Table 1). These rather simplistic simulations formed the basis for the Florida rule prohibiting aldicarb applications within 300 m of shallow drinking water wells on the Florida ridge.

Assessment

Sometimes, the objective of simulations is to obtain probability estimates for the occurrence of ground water concentrations underneath or downgradient

Table 1. Predicted Movement of Aldicarb Residues on the Florida Ridge

Water table depth (m)[a]	Saturated zone half-life (months)	Maximum distance (m) where aldicarb residue concentration exceeds 10 μg/l[b]	
		Single application	Yearly applications
1.5 (38)	6	120	132
	12	224	288
3.0 (31)	6	108	120
	12	216	270
7.2 (20)	6	78	102
	12	156	240

[a] Numbers in parentheses are the percent of applied aldicarb entering the saturated zone.
[b] Assumes an average ground water velocity of 0.15 m/day and an aldicarb application rate of 5.6 kg/ha (active ingredient). Values of other variables are given in Reference 20. The values for yearly applications are the asymptotes approached after a large number of years.

Adapted from Reference 20.

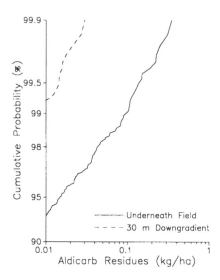

FIGURE 2   Cumulative probability distribution showing the predicted amount of aldicarb residues in shallow ground water underneath and 30 m downgradient of treated fields resulting from applications to North Carolina peanuts. (Adapted from Reference 20.)

of treated fields in a specific region such as a county or state. One approach[20] used a Monte Carlo procedure to estimate the probability of aldicarb residues underneath and downgradient of peanut fields in North Carolina (Figure 2). Probability density functions were developed for a number of variables, including soil type, soil properties as a function of depth for each soil type, weather data, degradation rates in the unsaturated and saturated zones, water table depth, and ground water velocity. Using these density functions, parameter sets were developed for 2000 cases. Each case was evaluated using a simple unsaturated zone model linked to PRZM. The maximum amounts of material (per hectare of aquifer) in the saturated zones underneath treated areas were calculated using a simple mass balance (adding the amount of incoming material to the previous amount of material less losses due to degradation) with a daily time step. Amounts of material at distances downgradient of the treatment area were calculated using the simple first-order rate equation where time was equal to the distance divided by the ground water velocity. The appropriateness of using such a simple model for this type of general assessment was confirmed by also evaluating several cases with a more comprehensive numerical model which included dispersion. This example demonstrates how even simplistic models can be extremely powerful tools in assessing the environmental fate of agricultural chemicals.

CONCLUSIONS

Saturated zone modeling, especially when combined with unsaturated zone modeling, can be an important tool in understanding the movement and degradation of agricultural chemicals in soil and ground water. Design of modeling simulations and the choice of models should be tailored to the study objectives. Complex models are often needed for site-specific interpretation of comprehensive field studies. However, simple models are often appropriate for assessment simulations performed to assist regulatory decisions or developing management practices.

REFERENCES

1. Zaki, M. H.; Moran, D.; and Harris, D. *Am. J. Public Health* 1982, 72, 1391–1395.
2. Cohen, D. B. In *Evaluation of Pesticides in Ground Water;* Garner, W. Y.; Honeycutt, R. C.; and Nigg, H. N., Eds.; ACS Symp. Ser. 315; American Chemical Society: Washington, DC, 1986; pp 499–529.
3. Ritter, W. F. *J. Environ. Sci. Health* 1990, B25(1), 1–29.
4. Jones, R. L. In *Hazardous Waste Containment and Treatment;* Cheremisinoff, P. N., Ed.; Encyclopedia of Environmental Technology, vol. 4; Gulf Publishing: Houston, TX, 1990; pp 355–376.
5. Bogardi, I.; Fried, J. J.; Frind, E.; Kelly, W. E.; and Rijtema, P. E. In *Proceedings of the International Symposium on Water Quality Modeling of Agricultural Non-Point Sources, Part 1;* DeCoursey, D. G., Ed.; ARS-81; U.S. Department of Agriculture, Agricultural Research Service: 1990, pp 227–252.
6. Duffy, C. J.; Kincaid, C. T.; and Huyakom, P. S. In *Proceedings of the International Symposium on Water Quality Modeling of Agricultural Non-Point Sources, Part 1;* DeCoursey, D. G., Ed.; ARS-81; U.S. Department of Agriculture, Agricultural Research Service: 1990, pp 253–278.

7. van der Heijde, P. K. M. and Prickett, T. A. In *Proceedings of the International Symposium on Water Quality Modeling of Agricultural Non-Point Sources, Part 1;* DeCoursey, D. G., Ed.; ARS-81; U.S. Department of Agriculture, Agricultural Research Service: 1990, pp 279–306.

8. Kinzelbach, W. K. H.; Dillon, P. J.; and Jensen, K. H. In *Proceedings of the International Symposium on Water Quality Modeling of Agricultural Non-Point Sources, Part 1;* DeCoursey, D. G., Ed.; ARS-81; U.S. Department of Agriculture, Agricultural Research Service: 1990, pp 307–325.

9. Knonikow, L. F. and Bredehoeft, J. D. *Computer Model of Two-dimensional Transport and Dispersion in Ground Water;* Techniques of Water Resources Investigations of the U.S. Geological Survey, Book 7, C2; U.S. Government Printing Office: Washington, DC, 1978.

10. Prickett, T. A.; Naymik, T. G.; and Lonnquist, C. G. *A Random-Walk Solute Transport Model for Selected Groundwater Quality Investigations;* Illinois State Water Survey Bulletin. 65; State of Illinois: Champaign, IL 1981.

11. Voss, C. I. SWTRA — Saturated-Unsaturated Transport — A Finite Element Simulation Model for Saturated-Unsaturated Fluid-Density-Dependent Groundwater Flow with Energy Transport or Chemically-Reactive Single-Species Solute Transport; Water Resources Investigations Report 84-4369; U.S. Geological Survey: Reston, VA, 1984.

12. Dean, J. D.; Huyakorn, P. S.; Donigian, A. S.; Voos, K. A.; Schanz, R. W.; Meeks, Y. J.; and Carsel, R. F. *Risk of Unsaturated/Saturated Transport and Transformation of Chemical Concentrations (RUSTIC).* Volume 1: *Theory and Code Verification;* EPA/600/3-89/048a; U.S. EPA Environmental Research Laboratory: Athens, GA, 1989.

13. Dean, J. D.; Huyakorn, P. S.; Donigian, A. S.; Voos, K. A.; Schanz, R. W.; and Carsel, R. F. *Risk of Unsaturated/Saturated Transport and Transformation of Chemical Concentrations (RUSTIC).* Volume 2: *User's Guide;* EPA/600/3-89/048b; U.S. EPA Environmental Research Laboratory: Athens, GA, 1989.

14. Huyakorn, P. S.; Kool, J. B.; and Wadsworth, T. D. In *Validation of Flow and Transport Models for the Unsaturated Zone: Conference Proceedings; May 23–26, 1988, Ruidoso, New Mexico;* Wierenga, P. J. and Bachelet, B. Eds.; Research Report 88-SS-04; Department of Agronomy and Horticulture, New Mexico State University: Las Cruces, NM, 1988, pp 176–186.

15. Jones, R. L.; Black, G. W.; and Estes, T. L. *Environ. Toxicol. Chem.* 1986, 5, 1027–1037.

16. Carsel, R. F.; Mulkey, L. A.; Lorber, M. N.; and Baskin, L. B. *Ecol. Modeling* 1985, 30, 49–69.

17. Carsel, R. F; Smith, C. N.; Mulkey, L. A.; Dean, J. D.; and Jowise, P. Users Manual for the Pesticide Root Zone Model (PRZM); EPA/600/3-84-109; U.S. EPA Environmental Research Laboratory: Athens, GA, 1984.

18. Wang, H. F. and Anderson, M. A. *Introduction to Groundwater Modeling, Finite Difference and Finite Element Methods;* W. H. Freeman: San Francisco, CA, 1982; pp 188–193.

19. Jones, R. L.; Hornsby, A. G.; Rao, P. S. C.; and Anderson, M. P. *J. Contam. Hydrol.* 1987, 1, 265–285.

20. Carsel, R. F.; Jones, R. L.; Hansen, J. L; Lamb, R. L.; and Anderson, M. P. *J. Contam. Hydrol.* 1988, 2, 125–138.

CHAPTER 8

Advances in Managing Agricultural Chemicals in Ground Water at the Farm Level

Russell L. Jones

Since the discovery of agricultural chemicals in drinking water wells in 1979, considerable progress has been made in understanding the movement and degradation of agricultural chemicals in the environment. Field and laboratory research studies have been designed and conducted, computer models have been developed, and numerous potable well monitoring studies have been performed. This research shows that residues in drinking water wells can be the result of movement of agricultural chemicals through the soil underneath treated fields or the result of spills or other point sources. Considerable progress has been made in developing management practices for protecting drinking water from both causes. Point source residues can be controlled using proper storage, mixing, loading, disposal, and chemigation practices. Chemical-specific management practices, such as optimizing application timing and well setback restrictions, can be used to protect drinking water supplies from non-point source residues. Proper siting and construction (including adequate casing depth and sealing) of drinking water wells can also reduce the potential for agricultural chemical residues.

In 1979 the detection of DBCP in drinking water wells in central California[1] and the detection of aldicarb residues in Long Island drinking water wells[2] demonstrated the potential for agricultural chemicals to move through the soil profile and into ground water. Prior to these detections, the effect that agricultural chemicals might have on ground water quality had been of little concern to most industry or regulatory scientists. Also, little research on the potential for agricultural chemicals to reach ground water had been conducted. The 1979 detections, along with other more recent instances of agricultural chemicals detected in drinking water wells,[3,4] sparked the initiation of a variety of monitoring programs, laboratory experiments, field research studies, and computer

0-87371-926-3/94/$0.00+$.50
© 1994 by CRC Press, Inc.

modeling research. The goals of these monitoring and research programs were to define the extent and magnitude of agricultural chemicals in drinking water wells and to develop management practices for minimizing such occurrences from future applications. The objective of this paper is to discuss the advances which have occurred in this area of research and the development of management practices which can be applied by farmers to protect drinking water wells in or near agricultural fields. Some recommendations for future research and regulations are also proposed.

Instances of agricultural chemicals in ground water are typically grouped into two categories. Non-point source is used to describe those instances where residues of agricultural chemicals enter ground water as a result of normal movement through the soil profile underneath a properly treated agricultural field. Point source is used to describe residues resulting from spills (often the result of improper storage, handling, mixing, equipment washing, or disposal practices), direct injection into a well during mixing or chemigation, or movement of surface water containing agricultural chemical residues down well casings or into sinkholes. Most instances where a number of drinking water wells in a region contain residues of a specific agricultural chemical (for example, DBCP in California, aldicarb in Long Island, or EDB in Florida) are the result of non-point sources. However, many of the isolated instances of agricultural chemical residues in drinking water wells (especially when high concentrations are observed) result from point sources. Often, determining whether residues in a drinking water well result from point or non-point sources is quite difficult since the well may be located near an agricultural field treated with the detected agricultural chemical and also near storage or mixing and loading areas. However, identification of the cause of an agricultural chemical detection in drinking water is a necessary prerequisite to the development of effective management practices. For example, if residues of an agricultural chemical are the result of rinsing application equipment near a well, then management practices such as optimizing the application timing to minimize movement through the soil profile will be ineffective.

ADVANCES IN RESEARCH

Since the discovery of agricultural chemicals in drinking water wells in 1979, considerable progress has been made on analyzing samples and on understanding how to conduct ground water-related research on agricultural chemicals.

Sample Analyses

Probably the most significant advance in ground water-related research has been the improvements in analyzing soil and water samples. The sensitivity of methods has generally increased 3 to 6 orders of magnitude during the last 10 to 20 years. Therefore, many of the instances of residues in drinking water could not have been detected at the time these products were registered. Analytical equipment has also been changing, with increasing use of HPLC (including post-column reactions) and mass spectroscopy in performing routine sample analyses. More selective detection systems have minimizing required cleanup steps, and combined with advances in instrumentation (such as autosamplers and electronic controls and recorders) have often reduced the amount of labor required per analysis. Also, the advances in analytical technology have made possible the increasing emphasis on the presence and behavior of metabolites. For example, until recently, soil samples from most field dissipation studies were analyzed for parent or occasionally one or two metabolites, while in recent studies analyses are performed for an increasing number of metabolites.

Potable Well Monitoring

During the past decade, monitoring programs have evolved into three types to meet distinctly different objectives. One type of monitoring program focuses on determining whether an agricultural chemical is appearing in drinking water wells in a certain region. Earlier programs with this objective often consisted of sampling a few of the most vulnerable wells in an area, while some more recent programs have allocated a large number of samples considering both vulnerability and product usage.[5] Well monitoring programs of the second type focus on identifying wells with residues after the potential for such residues has been identified in previous monitoring. Such monitoring programs, the design of which has changed little since 1979, usually consist of sampling a relatively large number of wells near treated fields. The third type of monitoring are statistically designed studies with the objective of assessing the exposure of a population to residues of an agricultural chemical. An example of an early study of this type is the aldicarb monitoring study conducted by the U.S. EPA on aldicarb in 1979. Examples of more recent studies are the National Pesticide Survey conducted by the EPA[6] and an alachlor study conducted by Monsanto (briefly reviewed in Reference 7). Such studies are relatively difficult to conduct since only a relatively small percent of drinking water wells in the U.S. contain agricultural chemical residues as demonstrated by the National Pesticide Survey. Recent studies have used stratified sampling designs to help focus on more vulnerable wells.

Field Research

Prior to 1979, most field research studies on environmental fate focused on the behavior of agricultural chemicals in surface soils. In field dissipation studies, samples of surface soils were collected to follow the persistence of an agricultural chemical usually to determine the effect on succeeding crops. Currently, one of the objectives of these studies in the U.S. is to identify compounds with potential to move to ground water. Sampling is to a depth of 0.9 m and if residues are found in samples in the deepest increment, ground water studies are usually required.

During the last decade, protocols for conducting comprehensive ground water studies have been developed.[5,8] Such studies consist of following the behavior of an agricultural chemical in both soil and ground water. Soil cores are used to follow the movement and degradation in the unsaturated zone, and shallow monitoring wells are installed and sampled to provide the same information for the saturated zone. In some studies, EPA requires the use of soil-suction lysimeters to provide a more sensitive indication of movement in subsoils. Advances in sampling techniques have contributed to the conduct of such studies. Manual procedures are now available for collecting soil samples as deep as 7.6 m[9] or installing wells in the upper 6 m of sandy aquifers when the water table is less than 7.6 m below the soil surface.[10] Drilling equipment is needed to collect deeper soil samples or install deeper wells. Because of the nature of the agricultural chemicals under study, a wider range of sampling equipment and techniques are often suitable compared to other situations where ground water monitoring must be performed (such as in hazardous waste cleanups).

Laboratory

Experiments performed in the laboratory have the advantage of usually being cheaper and more easily controlled than field experiments, Also, radio-labeled compounds can be used, simplifying analysis of samples and identification and quantification of important metabolites. However, during the past decade, researchers have shown that laboratory measurements are not always good indicators of behavior under field conditions. For example, laboratory measurements of degradation rates are usually performed at a constant temperature, while temperatures of surface soil in agricultural fields usually vary significantly during a 24-h period. Microbial populations may also be different under field and laboratory conditions.

Laboratory experiments concerning mobility of an agricultural chemical may also be misleading. Sorption is usually characterized by batch equilibrium measurements of the organic carbon/water partition coefficient, ignoring sorption rates and the hysteresis usually observed between sorption and desorption isotherms.[11] In experiments with repacked soil columns, the effects of soil structure on mobility (macropore flow through cracks or wormholes) cannot be simulated.

However, laboratory experiments can still be useful if their limitations are recognized by the researcher. Laboratory studies can help identify significant metabolites, the mechanism of degradation reactions (for example, whether a reaction pathway is chemical or microbial, or the influence of solid surfaces or soil constituents), provide relatively good indications of mobility, and determine the likely importance of sorption kinetics.

Field Lysimeters

Some researchers have tried to retain the advantages of experiments in soil columns while also conducting such experiments under field conditions (for example, References 12 and 13). In such experiments, undisturbed soil cores up to about a meter in length with a cross-sectional area of 0.02 to 1.0 m^2 of surface area are placed outdoors into the soil. In some facilities an underground room is constructed to provide access to the lysimeters for collecting water exiting the bottom of the lysimeters. In other installations the underground room is located several meters away to avoid any temperature distortions. In larger lysimeters, crops are often grown in and around the lysimeter to simulate field conditions better.

Lysimeter studies, like field dissipation studies, can identify those compounds in which some movement may occur below a depth of a meter from the soil surface. Some European countries such as Germany use lysimeter studies to determine the need for ground water studies, similar to the use of field dissipation studies in the U.S. In the author's opinion, the main advantage of lysimeter studies over field dissipation studies is the ability to use radio-labeled compounds. Therefore, lysimeters will become increasingly useful for exploring the behavior of new compounds applied at very low rates.

Modeling

During the past decade computer modeling has become an increasingly important tool in understanding and managing the movement of agricultural chemicals to ground water. Prior to 1979, the only existing non-point source models were on the movement of nitrates in soil. Currently, screening models are available for identifying agricultural chemicals which have the potential to move to ground water.[14,15] Some of the more widely used mechanistic models for simulating the behavior of agricultural chemicals in soil are RUSTIC[16,17] (the root zone module was previously issued as PRZM), GLEAMS,[18] CMLS,[19] LEACHMP,[20] MOUSE,[21] and RZWQM.[22] Other non-point models of interest include the stochastic model VULPEST[23] and the statistical model SESOIL.[24] Information on the use of these models is presented elsewhere.[25]

Comparisons of model simulations with field measurements have confirmed the usefulness of the models.[26-32] However, the limitations of the models must be understood by the modeler. None of the models do a good job of predicting soil concentration profiles,[27] but some do an adequate job of predicting the depth of residue movement or amount of movement below a specified depth.[28] By coupling an unsaturated zone model with a saturated zone model, the magnitude of any residues reaching ground water or the potential for residues in nearby drinking water wells can be evaluated.[28,29,33] The use of current modeling technology with existing meteorologic and soil data bases[34,35] permits essentially unlimited site-specific simulations. However, the accuracy of any modeling simulation is dependent upon the use of correct chemical, meteorological, soil, and crop data. Therefore, modeling is not a substitute for laboratory or field research, but rather a procedure for using such data in making predictions of agricultural chemical behavior in soil and ground water.

MANAGEMENT PRACTICES TO PREVENT POINT SOURCE RESIDUES

During the last decade, increasing emphasis has been placed on the proper use of agricultural chemicals to prevent point source contamination. Many programs have been initiated by state regulatory agencies, university extension agents, and agricultural chemical manufacturers to educate farmers and other agricultural chemical applicators on appropriate practices. Management practices to prevent point source residues are generic in nature and apply to all agricultural chemicals regardless of their chemical characteristics.

Many instances of point source residues can be attributed to spills near wells. Any spills which occur during the handling or transport of agricultural chemicals should be immediately cleaned up. Agricultural chemical manufacturers can provide advice for specific instances. Mixing and loading of agricultural chemicals should be performed in the fields to be treated, away from drinking water wells, or alternatively on a concrete pad so that any spills can be collected. Equipment should be washed away from any drinking water well. Rinsate from spray tanks should be collected and properly disposed of, preferably by spraying onto an appropriate crop. Empty containers of agricultural chemicals should be triple rinsed prior to disposal with the rinse water added to the spray tank.

Agricultural chemicals should be properly stored in a secured area to prevent deterioration of containers (for example, avoid storing cardboard boxes on the floor or other potentially moist locations). Purchases of agricultural chemicals by farmers should be planned to avoid storage for long periods of time. Plans should also be made on how to deal with a fire in a storage area. Water used to extinguish a fire can often create a greater problem than the fire itself.

Precautions must be taken to avoid directly introducing agricultural chemicals into wells. Open containers of agricultural chemicals should not be set on nearby wells during application (there have been several instances where the contents of these containers have been spilled directly into open wells). Also, mixing systems (both tank mixing and chemigation systems) should be designed to prevent water containing agricultural chemicals from back-siphoning into the well if the pump is shut off.

MANAGEMENT PRACTICES TO PREVENT NON-POINT SOURCE RESIDUES

Unlike management practices for preventing point source residues, management practices for preventing non-point source residues must be developed for a specific agricultural chemical, crop, and set of environmental conditions. Practices developed for one set of circumstances may not be applicable to a different set of circumstances. Management practices usually address normal variations in environmental parameters. Under abnormal conditions (for instance, heavy rains from a hurricane or a location where the velocity of ground water is high), management practices that are usually effective may be inadequate. The following paragraphs (adapted from Reference 36) illustrate the nature of possible management practices to minimize residues from non-point sources in ground water and protect drinking water supplies.

Application

One type of management practice is to optimize the application (timing, placement, and amount) of an agricultural chemical. For example, delaying aldicarb applications to potatoes from planting time to after plant emergence in the northeastern U.S. and Wisconsin[37,38] reduces the amount of aldicarb residues likely to reach ground water by avoiding heavy spring rains and by increasing degradation rates as a result of warmer soil temperatures. However, optimization of application timing must be tailored to local weather patterns. For example, delaying spring applications of aldicarb to Florida citrus until summer increases potential for movement of residues to ground water since rainfall is heaviest in the summer in central and southern Florida.[39]

The placement of an agricultural chemical may also be important. For some agricultural chemicals, the degradation rate may be faster if the material is placed near the soil surface rather than deeper into the soil.

Rate reductions usually have less effect than changes in other agricultural practices or differences in environmental conditions on the appearance or absence of residues. However, reducing the application rate to the minimum needed to provide pest control or the desired growth effect does contribute to reduced residues.

Irrigation

Irrigation practices can also affect the movement of agricultural chemicals. Because rainfall and irrigation water provide the driving force for downward movement of water (and therefore downward movement of agricultural chemicals), the amount of irrigation applied should be only that needed to compensate for evapotranspiration losses. Excessive irrigation will increase downward movement of water and, hence, agricultural chemicals.

The type of irrigation used may also affect potential movement of agricultural chemicals. For example, furrow irrigation may result in less movement than other irrigation techniques such as overhead sprinklers when agricultural chemicals are present only on plant foliage. For agricultural chemicals present in the soil, sprinkler irrigation often results in less movement because of the more even distribution of water throughout the field and because less water can be more easily applied during a single irrigation event. When irrigation water is applied unevenly to a field (such as in furrow irrigation), the placement of the agricultural chemical relative to the irrigation furrow may also influence downward movement.[28,40] For example, movement of an agricultural chemical applied to the bottom of an irrigation furrow will be greater than if the same chemical were applied to the soil between the irrigation furrows.

Well Buffer Zones

For those agricultural chemicals that continue to degrade upon reaching ground water, prohibiting applications within specified distances of drinking water wells may be an appropriate management practice for such agricultural chemicals. Such zones allow any residues that may enter ground water time to degrade before reaching nearby drinking water wells. The distances specified in these restrictions should be a function of both existing environmental factors and the characteristics of the agricultural chemical. Restrictions requiring buffer zones around shallow wells have been adopted for aldicarb applications in portions of Canada and the U.S. The basis for the distances specified in these restrictions has been a combination of field research, potable well monitoring data, and computer modeling.

Use Prohibitions

Under some circumstances, management practices such as those described previously may not provide adequate protection for underground drinking water supplies while maintaining the benefits of continued use of a specific agricultural chemical. In these circumstances, restrictions prohibiting applications of these agricultural chemicals may be necessary. The prohibition of the use of several different agricultural chemicals in the highly vulnerable region of Long Island, NY, is an example of such a restriction.

WELL LOCATION AND CONSTRUCTION REQUIREMENTS

Potable well monitoring and field research conducted over the past decade indicate that agricultural chemicals are more likely to occur in wells with certain characteristics. Such residues can be either point source or non-point source residues. Farmers or other people living near agricultural fields can minimize potential for residues of agricultural chemicals in drinking water by properly locating and constructing wells.

The location of drinking water wells affects potential for residues of agricultural chemicals, especially those resulting from point sources. Drinking water wells should be located away from areas used to store, load, mix, or dispose of chemicals. The well should also be located at least 15 m away from agricultural fields. This minimal buffer zone can help prevent damage to the well by application equipment, reduce potential for runoff water to transport agricultural chemical residues to the well, and provide some limited time for degradation and dispersion of any residues in ground water. Also, wells should not be located in depressions which collect runoff water.

Proper well construction is also important. Wells should be properly sealed at ground surface to prevent any runoff water from moving downwards along the well casing. Also, the area around the well casing should be properly sealed with grout or bentonite to prevent water near the top of the water table moving downwards along the well casing into the area of the aquifer near the well screen. Research studies indicate that most residues of agricultural chemicals which continue to degrade in ground water are usually contained in the upper 2 to 5 m of the saturated zone. Therefore, well casings should extend unbroken for at least 9 m below the water table.

RECOMMENDATIONS

In the author's opinion, the highest priority in ground water-related activities should be better use of already existing knowledge. This includes continuing education programs to help farmers properly store and use agricultural chemicals, conducting research and monitoring programs using existing protocols to obtain information on the environmental behavior of agricultural chemicals, and, when necessary, the development of appropriate management practices for minimizing potential for ground water residues and protecting drinking water wells.

Existing information suggests that some general regulatory actions could reduce instances of agricultural chemicals in drinking water wells. The application of any agricultural chemical within 15 m of any drinking water well should be prohibited. Regulations requiring adequate sealing and casing depth should be adopted for new drinking water wells, with appropriate timetables to upgrade existing drinking water wells (such regulations already exist in some locations). Since the potential for an agricultural chemical to move through the soil profile and into ground water is dependent on a variety of environmental parameters, regulators need to be receptive to more complex product labels and/or state management plans which can allow for product-specific and site-specific determination of required management practices (this is consistent with the recently issued U.S. EPA ground water strategy[41]).

Needs for future research include development of a relatively rapid laboratory procedure that can be used to estimate degradation rates under specific field conditions, better information on the variation of microbial degradation reactions with soil depth, and development of data bases indicating water table depth down to the field level. In addition, more basic research on water flow in soils (such as the effect of preferential flow through cracks, worm holes, or coarse sands) and sorption and desorption of chemicals onto and from soil should be continued.

REFERENCES

1. Cohen, D. B. In *Evaluation of Pesticides in Ground Water;* Garner, W. Y.; Honeycutt, R. C.; and Nigg, H. N., Eds.; ACS Symp. Ser. 315; American Chemical Society: Washington, DC, 1986; pp 499–529.
2. Zaki, M. H.; Moran, D.; and Harris, D. *Am. J. Public Health* 1982, 72, 1391–1395.
3. Holden, P. W. *Pesticides and Groundwater Quality, Issues and Problems in Four States;* National Academy Press: Washington, DC, 1986.
4. Ritter, W. F. *J. Environ. Sci. Health* 1990, B25(1), 1–29.
5. Jones, R. L. and Norris, F. A. In *Groundwater Residue Sampling Design;* Nash, R. G. and Leslie, A. R., Eds.; ACS Symp. Ser. 465; American Chemical Society: Washington, DC, 1991; pp 165–181.
6. *National Pesticide Survey Project Summary;* Office of Water, Office of Pesticides and Toxic Substances, U.S. EPA: Washington, DC, Fall, 1990.
7. Cohen, S. Z. *Ground Water Monit. Rev.* 1990, 10(4), 68.
8. Cohen, S. Z.; Eiden, C.; and Lorber, M. N. In *Evaluation of Pesticides in Ground Water;* Garner, W. Y.; Honeycutt, R. C.; and Nigg, H. N., Eds.; ACS Symp. Ser. 315; American Chemical Society: Washington, DC, 1986; pp 170–196.
9. Norris, F. A.; Jones, R. L.; Kirkland, S. D.; and Marquardt, T. E. In *Groundwater Residue Sampling Design;* Nash, R. G. and Leslie, A. R., Eds.; ACS Symp. Ser. 465; American Chemical Society: Washington, DC, 1991; pp 349–356.
10. Kirkland, S. D.; Jones, R. L.; and Norris, F. A. In *Groundwater Residue Sampling Design;* Nash, R. G. and Leslie, A. R., Eds.; ACS Symp. Ser. 465; American Chemical Society: Washington, DC, 1991; pp 214–221.
11. Wagenet, R. J. and Rao, P. S. C. In *Pesticides in the Soil Environment: Processes, Impacts, and Modeling;* Cheng, H. H., Ed.; Soil Science Society of America Book Series 2; Soil Science Society of America: Madison, WI, 1990; pp 351–399.
12. Steffens, W.; Fuhr, F.; and Mittelstaedt, W. In *Book of Abstracts Seventh International Congress of Pesticide Chemistry, Hamburg August 5–10, 1990;* Frehse, H.; Kesseler-Schmitz, E.; and Conway, S., Eds.; International Union of Pure and Applied Chemistry: Oxford, UK, 1990; Vol. III, p 78.
13. Smelt, J. H.; Schut, C. J.; and Leistra, M. *J. Environ. Sci. Health* 1983, B18, 645–665.
14. Gustafson, D. I. *Environ. Toxicol. Chem.* 1989, 9, 339–357.
15. Jury, W. A.; Focht, D. D.; and Farmer, W. J. *J. Environ. Qual.* 1987, 16, 422–428.

16. Dean, J. D.; Huyakorn, P. S.; Donigian, A. S.; Voos, K. A.; Schanz, R. W.; Meeks, Y. J.; and Carsel, R. F. *Risk of Unsaturated/Saturated Transport and Transformation of Chemical Concentrations (RUSTIC).* Volume 1: *Theory and Code Verification;* EPA/600/3-89/048a; U.S. EPA Environmental Research Laboratory: Athens, GA, 1989.

17. Dean, J. D.; Huyakorn, P. S.; Donigian, A. S.; Voos, K. A.; Schanz, R. W.; and Carsel, R. F. *Risk of Unsaturated/Saturated Transport and Transformation of Chemical Concentrations (RUSTIC).* Volume 2: *User's Guide;* EPA/600/3-89/048b; U.S. EPA Environmental Research Laboratory: Athens, GA, 1989.

18. Leonard, R. A., Knisel, W. A.; and Still, D. A. *Trans. ASAE* 1987, paper no. 86-2511.

19. Nofziger, D. L. and Hornsby, A. G. *Appl. Agricul. Res.* 1986, 1, 50–56.

20. Wagenet, R. J. and Hutson, J. L. *J. Environ. Qual.* 1986, 15, 315–322.

21. Steenhuis, T. S.; Pacenka, S.; and Porter, K. S. *Appl. Agricul. Res.* 1987, 2, 277–289.

22. DeCoursey, D. G.; Rojas, K. W.; and Ahuja, L. R. *Trans. ASAE* 1989, paper no. SW 89-2562.

23. Villeneuve, J.-P; Baton, O.; Lafrance, P.; and Campbell, P. G. C. In *Vulnerability of Soil and Groundwater to Pollutants, Proceedings and Information No. 38;* van Duijvenbooden, W. and van Waegeningh, H. G., Eds.; TNO Committee on Hydrological Research: The Hague, The Netherlands, 1987; pp 1097–1109.

24. Bonazountas, M. and Wagner, J. SESOIL: A Seasonal Soil Compartment Model; Office of Toxic Substances, U.S. EPA: Washington, DC, 1981.

25. Jones, R. L. In *Hazardous Waste Containment and Treatment;* Cheremisinoff, P. N., Ed.; Encyclopedia of Environmental Technology, vol. 4; Gulf Publishing: Houston, TX, 1990; pp 355–376.

26. Boesten, J. J. T. I.; van der Pas, L. J. T.; and Smelt, J. H. *Pest. Sci.* 1989, 25, 187–203.

27. Pennell, K. D.; Hornsby, A. G.; Jessup, R. E.; and Rao, P. S. C. *Water Resour. Res.* 1990, 26, 2679–2693.

28. Jones, R. L.; Black, G. W.; and Estes, T. L. *Environ. Toxicol. Chem.* 1986, 5, 1027–1037.

29. Jones, R. L.; Hornsby, A. G.; Rao, P. S. C.; and Anderson, M. P. *J. Contam. Hydrol.* 1987, 1, 265–285.

30. Carsel, R. F.; Mulkey, L. A.; Lorber, M. N.; and Baskin, L. B. *Ecol. Modeling* 1985, 30, 49–69.

31. Carsel, R. F.; Nixon, W. B.; and Ballantine, L. G. *Environ. Toxicol. Chem.* 1986, 5, 345–353.

32. Lorber, M. N. and Offutt, C. F. In *Evaluation of Pesticides in Ground Water;* Garner, W. Y.; Honeycutt, R. C.; and Nigg, H. N., Eds.; ACS Symp. Ser. 315; American Chemical Society: Washington, DC, 1986; pp 342–365.

33. Carsel, R. F.; Jones, R. L.; Hansen, J. L.; Lamb, R. L.; and Anderson, M. P. *J. Contam. Hydrol.* 1988, 2, 125–138.

34. Thompson, P. J.; Young, K.; Goran, W. D.; and Moy, A. *An Interactive Soils Information System User's Manual;* CERL Tr-N.87/18; U.S. Army Construction and Engineering Research Laboratory, Champaign, IL, 1987.

35. Carsel, R. F. and Jones, R. L. *Ground Water Monit. Rev.* 1990, 10(4), 96–101.

36. Jones, R. L. and Bostian, A. L. *Ground Water Monit. Rev.* 1989, 9(4), 75–78.
37. Wyman, J. A.; Jensen, J. O.; Curwen, D.; Jones, R. L; and Marquardt, T. E. *Environ. Toxicol. Chem.* 1985, 4, 641–651.
38. Jones, R. L.; Rourke, R. V.; and Hansen, J. L. *Environ. Toxicol. Chem.* 1986, 5, 167–173.
39. Jones, R. L.; Rao, P. S. C.; and Hornsby, A. G. In *Proceedings of Characterization and Monitoring of the Vadose (Unsaturated Zone), Las Vegas, December 8–10, 1983;* Nielson, D. M. and Curl, M., Eds.; National Water Well Association: Worthington, OH, 1984; pp 959–978.
40. Jones, R. L. *J. Contam. Hydrol.* 1987, 1, 287–298.
41. *Pesticides and Ground-Water Strategy;* 21T-1022, Office of Pesticides and Toxic Substances, U.S. EPA: Washington, DC, Oct., 1991.

CHAPTER 9

Industry's Perspective on Pesticide Issues Relating to Ground Water

Thomas J. Gilding

In agriculture, the issue of ground water protection has become a major subject of debate, one that is sure to influence decisions on how food and fiber will be grown in the future. The challenge of this debate is to identify and manage risks to ground water quality according to the *levels* of risks involved and their specific geographic *locations*. This must be done in ways that protect ground water, yet maintain our country's high level of agricultural productivity.

CHANGING AGRICULTURE

When addressing pesticide issues relating to ground water, it is important to look at them from a perspective of the overall "changes" going on in agriculture. One "change" I am specifically referring to has to do with the more recent emergence of environmental issues in agriculture. This environmental "change" can be characterized by an increased oversight and involvement on the part of the public in the way agriculture does business, including their expectations and demands for accountability.

As a result of a maturing public debate on agriculture and the environment, "sustainable agriculture" has become the term around which to craft future goals for agriculture. I believe a report prepared by the Council for Agricultural Science and Technology (CAST) provides a realistic perspective for defining goals for agriculture. This report[1] presents "long-term viability" of agriculture (sustainability) as having three distinct dimensions: (1) economic viability, (2) environmental and natural resources viability, and (3) social viability.

Indeed, economics, environmental protection, wise use of natural resources, and social values are all important goals for agriculture. I am one that believes that these different dimensions can be brought into agriculture in a way that maximizes achieving their respective goals. However, we cannot be so naive to think that during the process of developing and implementing policy, conflicts will not exist and trade-offs will not have to be made. The answers lie in a process of balancing these complementing, competing, and conflicting goals.

What this all means is that in defining and achieving production goals in agriculture, more attention needs to be given to assuring that natural resources and agricultural inputs, like pesticides, are used most efficiently, with minimum impact to the environment. The key here is *assuring* that natural resource or environmental problems are not occurring, and should they be occurring, necessary actions should be taken to correct them.

PESTICIDES AND GROUND WATER

The National Agricultural Chemicals Association (NACA) supports ground water policy that is focused on prevention of risk. The success in defining and implementing such policy, however, whether through legislation, regulations, or voluntary approaches, depends on the appropriate use of science-based risk assessment and management procedures and practices. It is essential, therefore, that ground water managers, both regulators and agricultural producers, have the necessary information to make informed and effective decisions.

NACA believes ground water protection programs for pesticides should be designed to: (1) protect public health and the environment; (2) minimize pesticide movement into ground water; and (3) maintain U.S. Agriculture's high level of productivity. NACA has adopted prevention-oriented goals in protecting ground water. This is achievable through the use of sound management practices by pesticide users in combination with the environmental stewardship programs of pesticide manufacturers.

Results from ground water monitoring studies, both on local and national scale, are showing that the presence of pesticide residues in ground water resources is not a widespread occurrence, either in numbers of pesticides being detected, or in their frequencies and levels of detection. What is shown, however, is that there are "problem areas," specifically being: (1) *localized geographic areas* where ground water is vulnerable, and (2) *improper practices* (human factors) resulting in spills during pesticide storage and mixing/loading, or pesticide wastes disposal. The key to effective protection of ground water from pesticides is to recognize where these "problem areas" exist and manage accordingly.

Providing pesticide users with the proper information to assure the safe and beneficial use of pesticides is the overall thrust of manufacturers' product stewardship programs. Although currently driven in response to environmental issues, these programs must also articulate principles of prudent use of pesticides. This would include using pesticides only when needed, in amounts necessary for what is intended, and in a manner that does not present unacceptable risk to health or the environment. The first two principles relate to maximizing cost effectiveness of pesticides, while the last principle addresses managing environmental and health risks.

In registering pesticides with the Environmental Protection Agency (EPA), manufacturers generate extensive information on the environmental and toxicological characteristics of their products. Part of this data requirement includes the properties of environmental fate, mobility, and degradation. In order for this information to be meaningful to pesticide users, specific use conditions and locations, relevant soil characteristics, and ground water vulnerabilities must be known. Perhaps assistance from local officials such as soil and water conservation districts could provide the necessary expertise for this need.

For manufacturers, the design of new pesticides with minimal leaching properties is frustrated by the fact that other criteria must also be considered in the registration process for pesticides. A few examples of these other criteria are solving the defined pest problem, avoiding health and safety concerns, and tailoring the product use to individual crop production systems. In addition, this research, development, and EPA registration process is time consuming and extremely costly.

It is estimated that as many as 10,000 individual chemicals fail a screening process for every product that has been successfully registered and commercialized. Deciding to commercialize a pesticide means that the manufacturer ultimately may have to commit to spend $30 to 40 million or more for research and development, not including costs to build new or remodel existing production facilities. The time from initial screening, through research, development, and final EPA approval normally takes up to 8 to 10 years.

It is also essential for pesticide manufacturers to work with pesticide users in defining sound management practices for managing those situations associated with handling, storage, mixing/loading, and disposal of pesticides which can present risks to ground water. The best approach for managing these risks is to *prevent* those situations which can lead to ground water contamination. Most notably this would involve preventing spills, protecting wells from mixing/loading activities, assuring integrity of wells themselves, and proper disposal of empty containers and application equipment rinse water.

NACA, with representatives from federal and state agencies, pesticide user groups, and researchers, has been actively involved in defining waste management strategies for empty pesticide containers and equipment rinse water. Written proceedings from two national conferences held in 1985 and 1986 were published by the EPA.[2] A final report on four regional workshops that followed in 1987 is also available.[3] This coordinated and cooperative initiative turned out to be a very positive step towards solving pesticide users' waste disposal difficulties.

ROLE OF TECHNOLOGY

Normal evolution of technology and practices over the last 40 years has brought us to our current enviable level of agricultural efficiency and productivity. Agricultural pesticides have played an important part in this accomplishment and will continue to be a *positive* factor in the economic dimension of a sustainable agriculture. Farmers, based on *their* experience, use pesticides as a cost effective means in helping to maximize efficiency and productivity in crop yields and in protecting those yields once achieved from risks of insects or disease.

Certainly, future technology advances will continue to make important progress in pest control strategies by: (1) minimizing or eliminating pest risks in the first place, and (2) improving pest control strategies for those situations requiring such measures. The elimination of pest threat is obviously the most ideal situation for farmers and should be a major goal of technology. However, we do live in a world where things not planned or wanted happen. Threats from pests to our food and fiber production are an example of "not planned or wanted" and, unfortunately, will be around for some time in the foreseeable future.

As we strive to further improve agriculture's ability to effectively control pest problems, standardized criteria is needed for guiding and selecting pest control strategies (chemical or non-chemical). In referencing "sustainable agriculture" as defined earlier, pest control strategies should attempt to balance cost effectiveness and environmental risks. The optimum being to *maximize* cost effectiveness in controlling pests, while *minimizing* environmental risks in doing so.

Factors which contribute to the potential for a pesticide to get into ground water resources from its actual registered use can be grouped by the following three distinct, but related categories. These are (1) properties of the pesticide; (2) conditions of the application site; and (3) methods and timing of application. Within these categories, there are specific properties or conditions, each having some level of effect on the "potential" for a pesticide to get (or *not* get) into ground water. In assessing potential risks to ground water for a given pesticide at a given application site, these categories need to be considered in combination and prioritized. The important factors within each category are

- *Pesticide properties* — "Pesticides" are chemicals which have a wide range in potential to move into water resources. The two most important properties of a pesticide chemical affecting its potential to get into ground or surface water, all other things being equal, are the rate of degradation (persistence) and ability to adsorb to organic content in the soil or to soil particles.
- *Site conditions* — Vulnerability of ground or surface water at the application site is primarily determined by the soil present, distance to ground or surface water, and climatic factors (amount and intensity of rainfall or irrigation). The soil characteristics of most importance are the texture, clay content, organic content, and water holding capacity.

- *Methods and timing of application* — How a pesticide is applied (from both placement and loads) and when (relative to rain events or irrigation) can have significant effects on its potential to eventually get into ground or surface water.

Technology holds great promise for assessing and managing pesticide risks to ground water. What is involved is using our knowledge and understanding to make the most realistic simulation of the leaching characteristics of pesticides under specific application conditions. This can be done through simple numerical models or more complex mechanistic models with the aid of computers. Within computer models, there can be a wide range in sophistication and applications. There are some basic models that only have value as educational tools, while more complex models can be used to evaluate alternative management practices or make regulatory risk assessments. Models are also being used by manufacturers in their research and development programs on new products.

SUMMARY

The agricultural chemicals industry is committed to the protection of ground water quality. We recognize this is a goal that we all share. Likewise, preserving our nation's agricultural productivity is a goal that we all share. These parallel goals in the sustainability of agriculture are complex, but achievable, provided that the problems and their solutions are clearly defined along with commitment and responsible action.

A statement in a video program released by the American Soybean Association and the National Corn Growers Association[4] summarizes to the point what needs to be done in our search for sustainability in agriculture. The statement is "in balancing the parallel needs for protecting ground water and preserving agricultural productivity, it is important that the agricultural community recognizes that this is simply not a productivity issue. The millions of people that are served by the bounty of America's farms must recognize that it is simply not an environmental issue. The best interests of all parties are served when ground water is aggressively protected and agricultural productivity is maintained."

REFERENCES

1. *Long-Term Viability of U.S. Agriculture;* Council for Agricultural Science and Technology; Report No. 114; June 1988.
2. *EPA Proceedings: National Workshop on Pesticide Waste Disposal*, September 1985 (Doc # EPA/600/9-85/030) and September 1987 (Doc # EPA/600/9-87/001).

3. *Managing Pesticide Wastes: Recommendations for Action;* National Agricultural Chemicals Association, 1988.
4. *Ground Water and Agricultural Chemicals: Understanding the Issues,* American Soybean Association and the National Corn Growers Association, 1988.

CHAPTER 10

National Pesticide Survey: Methods, Results, and Policy Implications

Jeanne S. Briskin

The U.S. Environmental Protection Agency conducted a national survey to learn about the prevalence of 126 pesticides and pesticide degradates plus nitrates in public and private drinking water wells. This paper describes the statistical design and chemical methods used in the survey. Then, estimates of contamination rates for public and domestic water supply wells are presented. Progress regarding the investigation of the relationship among contamination patterns, pesticide use, and ground water vulnerability are reported. Finally, possible uses of the survey results in the Federal drinking water and pesticides programs are discussed.

INTRODUCTION

The U.S. Environmental Protection Agency (EPA) recently completed a national survey of pesticides and nitrates in public and private drinking water wells. Designed in 1985, the survey had two goals: (1) to determine the frequency and concentration of pesticides and nitrate in drinking water wells nationwide, and (2) to improve EPA's understanding of how the presence of pesticides and nitrate in drinking water wells is associated with patterns of pesticide use and the vulnerability of ground water to contamination. The Phase I report released in November 1990 meets the first of these goals. This paper describes the stratified design, and chemical analysis methods that were used by the survey. Then the results of the Phase I analysis are presented.

Table 1. Agricultural Pesticide Use

Level	Percent of total county area with agricultural pesticides applied
High	≥75
Moderate	30–74
Low	5–29
Uncommon	<5

Adapted from Reference 1.

Finally, the paper discusses how EPA will use the results of the survey in its pesticide, ground water, drinking water, and non-point source water pollution programs.

SURVEY DESIGN

The National Pesticide Survey[1] was designed to yield results that are statistically representative of the nation's 94,600 community water supply (CWS) wells and 10.5 million rural domestic water supply (DWS) wells. To choose a sample of wells for testing, the EPA characterized all counties in the U.S. according to pesticide use and the relative vulnerability of the ground water to contamination — two critical factors affecting the presence of pesticides in drinking water wells. For the first category, the EPA concentrated on agricultural pesticide use, specified as high, moderate, low, or uncommon. Because actual county level data on pesticide use were not available, the proportion of the county land area used to grow crops that rely on pesticides of interest to the survey was used as a surrogate. Table 1 summarizes the definitions for this stratification variable.

For the second stratification category, the EPA used a numerical classification system called DRASTIC to roughly group counties into categories for high, medium, or low vulnerability to ground water contamination. DRASTIC considers seven factors: depth of water table, recharge rate of the aquifer, aquifer media, soil type, topography, impact of the unsaturated zone, and the hydraulic conductivity of the aquifer. These factors were assessed in a general way for each county, and counties were assigned to a high, medium, or low vulnerability category based on their overall average score. Table 2 summarizes the definitions for the vulnerability categories.

In all, 12 strata (4 pesticide use categories times 3 vulnerability categories) were established. Counties were randomly chosen from each strata, with a slight over-representation from the higher risk categories. (Therefore, the strata carried unequal weights which were accounted for in the calculation of results.) Then wells were identified using a master list of community water system wells and a random digit dialing survey within sampled counties for rural domestic wells. Finally, wells that met survey eligibility criteria were selected randomly from each county. Table 3 lists the final sample sizes for each stratum. The

Table 2. Ground Water Vulnerability

Agricultural pesticide use	Ground water vulnerability[a]	Defining DRASTIC scores
High	High	≥148
	Moderate	116–147
	Low	≤115
Moderate	High	≥163
	Moderate	113–162
	Low	≤112
Low	High	≥159
	Moderate	132–158
	Low	≤131
Uncommon	High	≥152
	Moderate	121–151
	Low	≤120

[a] High = 75th percentile value for the given pesticide use category. Low = 25th percentile value for the given pesticide use category.

Adapted from Reference 1.

Table 3. National Pesticide Survey Sample Size

Stratum	Agricultural pesticide use	DRASTIC score	Design DWS	Sample Size[a] CWS
1	High	High	91	27
2		Moderate	93	26
3		Low	48	14
4	Moderate	High	50	45
5		Moderate	72	32
6		Low	31	25
7	Low	High	77	73
8		Moderate	112	68
9		Low	53	53
10	Uncommon	High	67	71
11		Moderate	88	90
12		Low	45	75
			827	599

[a] Number of systems selected, including allowances for cancellations, refusals to participate, and inability to sample prior to treatment. Minimum desired sample size after dropout was 734 DWS wells and 564 CWS wells.

Adapted from Reference 1.

EPA actually sampled 540 community water supply wells in all 50 states, and sampled 752 rural domestic wells in 90 counties in 38 states.

A series of seven questionnaires were developed to gather information on the eligibility of wells for testing and on pesticide use, land use, well construction, farming practices, and other factors which might affect the presence and levels of pesticides in drinking water wells. The questionnaires were coded to allow easy cross-checking for internal consistency. Space for drawing a map of the area surrounding each well and a diagram to describe pertinent treatment and other physical factors for later qualitative analysis were also included.

ANALYTICAL CHEMISTRY

The selected 101 pesticides, 25 pesticide by-products or degradates, and nitrate for chemical analysis. Table 4 lists the chemicals measured in the survey. The analytes were chosen based on their previous occurrence in ground water, high volume of use, and potential to contaminate drinking water wells (long half-life and/or high water solubility). Additional chemicals were analyzed if the chosen multi-residue methods would also identify them and measure their concentrations.

Six new multi-residue methods were developed for the survey, and two existing single residue methods were also used. Table 4 also lists these methods. Potential analytes were omitted from the methods if serious degradation during sample storage was demonstrated during analyte stability studies. The EPA worked with the Association of Analytical Chemists to conduct a multi-laboratory validation study of the new methods. The multi-laboratory validation study was completed in late 1990, and the methods were found to perform well. The EPA formally adopted the methods for enforcement purposes, and requires that they be used by public water suppliers in order to comply with federal drinking water regulations.

Chemical analyses of NPS water samples were performed at five contract laboratories and three EPA referee laboratories. The contract laboratories were responsible for the primary analyses. The referee laboratories performed quality assurance analyses and provided additional backup capacity to support the primary, contract laboratories.

For 112 of the 127 analytes, minimum reporting limits were established, above which the analyte was reported as having been detected. Minimum quantification limits are levels above which the concentration as well as detection was reported. For 15 analytes which were either unstable or lacked reliable method performance, no reporting limit was established. The presence of these analytes is reported as a positive detection without quantification.

Positive detections were confirmed by reanalysis of all suspected positive extracts using a second (different) gas chromatographic or high performance liquid chromatographic column. Analytes detected by gas chromatography were also confirmed using mass spectroscopy. Depending on the method, the EPA found that between 22 and 47% of suspected detections were not confirmed by mass spectroscopy. This result emphasizes the importance of confirming analyses to avoid false positives when identifying contaminants of drinking water. The results of the primary column were reported as the quantitative results when the secondary column agreed with the primary results to within 25%. If this requirement was not met, and the presence of the analyte was confirmed by mass spectroscopy, then the detection was reported without quantification.

Potential problems with false negatives were assessed by comparing the results from the primary laboratories with those from the referee laboratories.

Table 4. Survey Analytes and Analytic Methods

NPS METHOD 1: Gas Chromatography with a Nitrogen-Phosphorous Detector (45 Analytes)

Alachlor	Diphenamid	Methyl Paraoxon	Simazine
Ametryn	Disulfoton[a]	Metclachlor	Simatryn
Atraton	Disulfoton sulfone[a]	Metribuzin	Stirofos
Atrazine	Disulfoton sulfoxide[a]	Mevinphos	Tebuthluron
Bromacil	EPTC	Molinate	Terbacil
Butachlor	Ethoprop	Napropamide	Terbufos*
Butylate	Fenamiphos	Norflurazon	Terbutryn
Carboxin	Fenarimol	Pebulate	Triademefon
Chlorpropham	Fluridone	Prometon	Tricyclazole
Cycloate	Hexazinone	Prometryn	Vernolate
Diazinon*	MGK 264	Pronamide*	
Dichlorvos	Merphos*	Propazine	

NPS METHOD 2: Gas Chromatography with an Electron Capture Detector (29 Analytes)

4,4-DDD	Dielotrin	Heptachlor epoxide	gamma-HCH
4,4-DDE	Endosulfan I	Hexachlorobenzene	alpha-Chlordane
4,4-DDT	Endosulfan II	Methoxychlor	gamma-Chlordane
Aldrin	Endosulfan sulfate	Propachlor	cis-Permethrin
Chlorobenzilate[a]	Endrin	Trifluralin	trans-Permethrin
Chloroneb	Endrin aldehyde	alpha-HCH	
Chlorothalonil	Etridiazole	beta-HCH	
DCPA	Heptachlor	delta-HCH*	

NPS METHOD 3: Gas Chromatography with an Electron Capture Detector (17 Analytes)

2,4-D	4-Nitrophenol[a]	Dalapon[a]	Pentachlorophenol (PCP)
2,4-DB	Acifluorfan[a]	Dicamba	Picloram
2,4,5-TP	Bentazon	Dicamba 5-hydroxy-	
2,4,5-T	Chloramben[a]	Dichlorprop	
3,5-Dichlorobenzoic acid	DCPA acid metabolites	Dinoseb	

Table 4. Survey Analytes and Analytic Methods (continued)

NPS METHOD 4: High-Performance Liquid Chromatography with an Ultraviolet Detector   (18 Analytes)

Atrazine, deethylated	Diuron	Metribuzin DA	Propanil
Barban	Fenamiphos sulfone	Metribuzin DADK[a]	Propham
Carbofuran, phenol-3-keto-	Fenamiphos sulfoxide	Metribuzin DK[a]	Swep
Carbofuran, phenol	Fluometuron	Neburon	
Cyanazine	Linuron	Pronamide metabolite	

NPS METHOD 5: Direct Aqueous Injection HPLC with Post-Column Derivatization   (10 Analytes)

Aldicarb	Baygon	Carbofuran, 3-hydroxy-	Oxamyl
Aldicarb sulfone	Carbaryl	Methiocarb	
Aldicarb sulfoxide	Carbofuran	Methomyl	

NPS METHOD 6: Gas Chromatography with a Nitorgen-Phosphorous Detector   (1 Analyte)

Ethylene thiourea (ETU)

NPS METHOD 7: Microextraction and Gas Chromatography   (5 Analytes)

Ethylene dibromide (EDB)	1,2-dichloropropane[b]	trans-1,3-dichloropropene[b]
Dibromochloropropane (DBCP)	cis-1,3-dichloropropene[b]	

NPS METHOD 9: Automated Cadmium Reduction and Colorimetric Detection   (1 Analyte)

Nitrate and nitrite measured as nitrogen (N)

a   Qualitative only.
b   Method 8 dropped. Analytes previously included in Method 8 also detectable by Method 7.

From National Survey of Pesticides in Drinking Water Wells — Phase I Report, p. 39.

Taking differences between the minimum quantification limits among laboratories into account, the referee laboratories detected an analyte that was not found by the primary laboratories in 2 cases out of 200 samples. Therefore, EPA estimates that the false negative rate was about 1% or less for each analyte.

Precision and accuracy were determined at the primary laboratories using spiked samples. Comparison of the results at primary laboratories to laboratory control standards indicates that the sample matrix had virtually no impact on analytical precision and accuracy. The EPA also found that similar precision and accuracy were obtained by both the primary and referee laboratories.

QUALITY ASSURANCE/QUALITY CONTROL

Comprehensive quality assurance (QA) activities were managed by a full-time QA officer. Leaders of all parts of the project were directly responsible for assuring that standard procedures were followed and that the information was gathered and analyzed in an accurate, timely manner. Quality control (QC) activities included independent review of the survey design, training for samplers and other field survey personnel, and audits of field and laboratory activities to ensure that data was collected, and day-to-day monitoring of well sampling and laboratory analysis was completed in a careful and accurate manner. Quality control measures were also followed during sample collection and shipping. The EPA conducted performance evaluation studies of the analytical laboratories to monitor the accuracy and precision of the chemical analyses. Special computerized logic checks were completed to assure that data entry was accurate and that information provided on the seven questionnaires from each well were reasonable and internally consistent.

Several steps to ensure quality in the laboratories were undertaken. Verified calibration standards were provided by the EPA to all participating laboratories. Laboratory control standards were analyzed to ensure consistent, high recovery of analytes during extraction processes. Control charts traced the recovery of laboratory control standards and surrogate standards. Instrument control standards were used each day for NPS methods 1 to 6 to ensure that instruments were properly calibrated and functioning prior to sample analysis. Surrogate standards (methods 1 to 7) were used to demonstrate proper analyte recovery in individual samples. Method blanks and shipping blanks were analyzed to control false positives.

RESULTS OF PHASE I ANALYSIS

The Phase I report of results concentrates on national estimates of the number of community water supply wells and rural domestic water supply wells that contain any of the 127 pesticides, pesticide degradates, or nitrates analyzed in the survey.

Based on the survey results, the EPA estimates that 9850 (10.4%) community water supply (CWS) wells and 446,000 (4.2%) rural domestic water supply (DWS) wells contain detectable levels of one or more pesticides. Fewer than one percent of the drinking water wells are expected to contain one or more pesticides at levels of health concern — EPA health advisory levels or Maximum Contaminant Levels. This corresponds to ≤750 CWS wells and 60,900 DWS wells.

More than half of all drinking water wells (49,300 CWS and 6 million DWS) are estimated to contain detectable levels of nitrate. There are 1130 CWS wells (1.2%) and 254,000 DWS wells (2.4%) that exceed the maximum contaminant level for nitrate of 10 mg/l nitrate (measured as nitrogen). Tables 5 and 6 present the corresponding 95% confidence intervals for these estimates.

The most commonly detected pesticide product was a degradate of dacthal or DCPA. EPA estimates that about 6010 CWS wells (6.4%) and 264,000 DWS wells (2.5%) contain detectable levels of DCPA acid metabolites. The health advisory level for dacthal and its metabolites is 4000 mg/l, or about 1000 times greater than the detection and quantification limit. No wells sampled in the survey were found to exceed this health-based limit. Dacthal is an herbicide once used primarily on lawns, turf, and golf courses. Now, the major use of dacthal is on fruits and vegetables. Dacthal degrades relatively quick and tends to be immobile in soil, but the products of degradation, the mono- and di-acid forms, are both mobile and persistent.

The second most frequently found pesticide was atrazine, an herbicide used on corn and sorghum. The EPA estimates that 1570 CWS wells (1.7%) and 70,800 DWS wells (0.7%) contain detectable levels of atrazine. Other pesticides and degradates detected in the survey include simazine, prometon, hexachlorobenzene, dibromochloropropane (DBCP), dinoseb, ethylene dibromide (EDB), lindane, bentazon, ethylene thiourea (ETU, a degradate of EBDC fungicides), alachlor, chlordane and 4-nitrophenol, a degradate of parathion. Five pesticides — alachlor, atrazine, DBCP, EDB, and lindane — were detected above health based limits in DWS wells. Table 5 presents estimates of the number of wells that are estimated to contain detectable levels of these pesticides, together with the 95% confidence intervals.

Based on the precision of the survey, the EPA estimates that the maximum number of wells that may contain pesticides which the survey analyzed for but did not detect is 750 CWS wells and 83,100 DWS wells. This upper 95% confidence bound corresponds to 0.8% of the wells.

PHASE II ANALYSIS

Most policymakers at the EPA and the U.S. Department of Agriculture are interested in analyses of possible relationships among contamination of wells,

Table 5a. Estimated Number and Percent of Community Water System Wells[a] Containing NPS Analytes[b]

Analyte	Estimated Number	95% Confidence Interval (lower-upper)	Estimated Percent	95% Confidence Interval (lower-upper)
Nitrate	49,300	(45,300–53,300)	52.1	(50.0–56.3)
DCPA acid metabolites	6,010	(3,170–8,840)	6.4	(3.4–9.3)
Atrazine	1,570	(420–2,710)	1.7	(0.5–2.9)
Simazine	1,080	(350–2,540)	1.1	(0.4–2.7)
Prometon	520	(78–1,710)	0.5	(0.1–1.8)
Hexechloro-benzene[c]	470	(61–1,630)	0.5	(0.1–1.7)
Dibromochloro-propane (DBCP)[c]	370	(33–1,480)	0.4	(<0.1–1.6)
Dinoseb[c]	25	(1–870)	<0.1	(<0.1–0.9)

Table 5b. Estimated Number and Percent of Rural Domestic Wells[d] Containing NPS Analytes

Analyte	Estimated Number	95% Confidence Interval (lower-upper)	Estimated Percent	95% Confidence Interval (lower-upper)
Nitrate	5,990,000	(5,280,000–6,700,000)	57.0	(50.3–63.8)
DCPA acid metabolites	264,000	(129,000–477,000)	2.5	(1.2–4.5)
Atrazine	70,800	(13,300–214,000)	0.7	(0.1–2.0)
Dibromochloro-propane (DBCP)[c]	38,400	(2,740–164.000)	0.4	(<0.1–1.6)
Prometon	25,600	(640–142,000)	0.2	(<0.1–1.4)
Simazine	25,100	(590–141,000)	0.2	(<0.1–1.3)
Ethylene dibro-mide[c]	19,200	(160–131,000)	0.2	(<0.1–1.2)
gamma-HCH	13,100	(14–120,000)	0.1	(<0.1–1.1)
Ethylene thiourea	8,470	(1–111,000)	0.1	(<0.1–1.1)
Bentazon	7,160	(1–109,000)	0.1	(<0.1–1.0)
Alachlor	3,140	(1–101,000)	<0.1	(<0.1–1.0)

a   Total number of community water system wells in the United States is equal to 94,600.
b   No estimates possible for alpha-chlordane, gamma-chlordane, beta-HCH, and 4-nitrophenol.
c   Registration cancelled by EPA.
d   Total number of rural domestic wells in the United States is equal to 10,500,000.

From National Survey of Pesticides in Drinking Water Wells — Phase I Report, p. 66.

pesticide use, agricultural practices, ground water vulnerability, well construction, and other factors. The EPA will use information from other studies to formulate specific hypotheses to test during the Phase II analysis. We hope to identify practices which affect ground water quality and factors which can predict areas at greatest risk for contamination of ground water by pesticides and nitrates.

Phase II will investigate some of the unexpected findings from the national projections provided in Phase I. In particular, we were surprised to learn that CWS wells were more likely to contain pesticides than DWS wells. We were also surprised to discover that the degradates of dacthal were the most commonly found pesticide. We plan to use regression analysis and other methods to investigate possible factors which might explain these patterns.

Temporal or seasonal factors may have affected the survey's results. We sampled each well one time, and did not revisit wells in later seasons. Because sampling lasted more than 1 year, we obtained more samples during fall and winter than during the spring and summer months. In addition, many areas of the U.S. were suffering from a drought during the survey's sampling period.

Although the survey did not collect data on amount or timing of rainfall compared to dates of pesticide application or water sample collection, we have proposed several investigations to assess the importance of temporal variability and seasonal trends on the study results. First, we checked to see if the excess of fall and winter samples were also lumped geographically. We found that during each season, we collected approximately the same number of samples from each region of the country; we avoided taking all the spring samples in Montana and all the winter samples in Florida. Second, we gathered data from the National Weather Service on rainfall and air temperature to compare to the sampling dates and locations from the survey. Through exploratory analysis, we may be able to discern some trends relating weather to drinking water well quality.

An area of continuing interest is national exposure to pesticide and nitrates in drinking water. Exposure and risk estimates require additional assumptions and calculations from those completed in the Phase I NPS report. In the Phase II report, the EPA will attempt to estimate the distribution of concentrations of pesticides and nitrates detected by the survey. An assessment of national exposure to these chemicals will then be developed. This exposure estimate may be useful in assessing health risks.

HOW THE EPA WILL USE THE NPS RESULTS

EPA will use the NPS results to improve policies, guidance, and regulations developed under the Safe Drinking Water Act and the Federal Insecticide, Fungicide, and Rodenticide Act. As required by the Safe Drinking Water Act, the EPA published a list of candidate chemicals for possible future regulation. Included in this list were 14 pesticides analyzed by the EPA in the NPS that had not previously been scheduled for review or regulation.[2] EPA promulgated 15 new Maximum Contaminant Levels (MCLs) for pesticides analyzed by the survey, and monitoring requirements for 19 unregulated pesticides and degradates.[3] The EPA plans to establish MCLs for a number of additional pesticides detected by the survey within the next year or two.

Water suppliers must monitor their water periodically to assure that the water they deliver meets state and federal standards. Supplies which have previously detected contaminants or are vulnerable to contamination must be monitored more frequently than protected, pristine supplies. The EPA plans to publish guidance for state regulatory authorities and water suppliers. The guidance will assist water suppliers and state regulators in assessing the vulnerability of the systems to contamination. The EPA will use the Phase II analysis to prepare more practical, effective guidance for this program.

The Phase II analysis will also assist the EPA in identifying conditions under which wellhead protection programs are needed. Under the SDWA, these programs are designed to prevent contamination of ground water, particularly in recharge areas which serve public water supplies. The EPA will improve the technical assistance it provides to state and local governments by using the NPS results.

Under FIFRA, the EPA may impose limits on the use of pesticides in order to protect drinking water supplies. These limits appear on the labels of the pesticides. The NPS results will help the EPA to identify chemicals and pesticide use practices that need label changes to assure that future pesticide use does not endanger drinking water wells. An additional condition on pesticide use that the EPA may impose is restricted use. Restricted use pesticides may only be used by certified applicators. The EPA will consider which pesticides detected, if any, should be handled only by certified applicators who receive special training on how to protect ground water.

The EPA is planning to establish a Pesticide in Ground Water Strategy in June 1991. Under this strategy, states would establish pesticide management plans designed to protect ground water from contamination. The EPA is working with the states to write guidance on how to identify areas which are vulnerable to and practices which can prevent contamination. The Phase II report should identify key factors associated with higher risks of pesticides in drinking water wells which can be highlighted in the state management plans and guidance.

The widespread occurrence of nitrate in drinking water wells, and the large number of public and private water supplies that exceed the MCL for nitrate, are of great concern to the EPA. In response to the survey results, the EPA established a workgroup to determine what activities the EPA or others could take to protect ground and surface water resources from nitrate contamination. To accomplish the goals of protecting drinking water supplies and preventing additional pollution, the workgroup recommended five broad areas for action: improved enforcement of the drinking water standard, improved control of point sources of pollution, adoption of better nutrient management systems, improved fertilizer use, and research to allow resolution of unclear areas. In 1991, the EPA plans to work with the USDA and others to identify and implement specific activities to accomplish these goals.

The views expressed in this paper are those of the author, not necessarily those of the U.S. Environmental Protection Agency.

LITERATURE CITED

1. *National Survey of Pesticides in Drinking Water Wells: Phase I Report*, EPA 570/9-90-015, U.S. Environmental Protection Agency, Washington, DC, 1990.
2. 56 *Federal Register* 1470–1474, 1991.
3. 56 *Federal Register* 3526–3597, 1991.

CHAPTER 11

Assessing Leaching Potential in California Under the Pesticide Contamination Prevention Act

Bruce R. Johnson, J. T. Leffingwell, and M. Rose Wilkerson

The Pesticide Contamination Prevention Act (PCPA) of 1985 in California mandated a framework for the assessment of pesticide leaching potential. That framework included a requirement for registrants to provide chemical and environmental fate data and for the California Department of Food and Agriculture (CDFA) to calculate Specific Numerical Values to quantify cut-off points for use in determining leaching potential. As of April 1991 approximately 6100 studies had been submitted by registrants to the CDFA to satisfy the data requirements. Specific Numerical Values have been established for solubility, soil adsorption, hydrolysis, and aerobic and anaerobic soil metabolism. These values are used as a first screen to identify pesticides with leaching potential. A second screen is used to classify compounds according to whether or not they are soil applied. The resulting list of pesticides is the Ground Water Protection List and determines CDFA priorities for ground water monitoring.

In 1985 the PCPA[1] set in motion a mechanism which profoundly influenced pesticide regulation in California. In this paper we provide a brief summary of the history of the Act and describe how leaching potential assessment is conducted under the Act.

THE PESTICIDE CONTAMINATION PREVENTION ACT

By 1985 there had been a proliferation of reports of pesticide detection in ground water. Findings of herbicides in the Midwest,[2,3] aldicarb in New York and Florida,[4,5] and various pesticides in California[6] prompted several studies of

Table 1. Statistics on Acceptable Tests by Test Type Received by CDFA as of April 1991

Test	Total texts submitted	Original test reports submitted	Acceptable original test reports	Percent acceptable test reports
Solubility	537	409	178	44
Vapor pressure	540	392	177	45
Henry's Law constant	444	369	129	35
Octanol/water coeff.	490	391	192	49
Soil adsorption coeff.	719	597	195	33
Hydrolysis	433	353	186	53
Water photolysis	457	360	159	44
Soil photolysis	385	286	148	52
Aerobic soil metab.	642	483	155	32
Anaerobic soil metab.	487	323	114	35
Field dissipation	994	782	108	14
Total	6128	4745	1741	

the problem.[7-9] In response to these developments, the California Legislature adopted legislation "... to prevent further pesticide pollution of the ground water aquifers of this state" (Reference 1:Section 13141(g)). The resulting law, the PCPA of 1985 (or the Act), required CDFA to ask registrants for specific environmental fate data, use these data to identify pesticides with leaching potential, monitor ground water and soil for such pesticides, maintain a data base of wells sampled for pesticide residues, and follow a specified procedure to review and modify use of pesticides found in ground water due to agricultural use.

The PCPA had several important features, which have had a profound impact on its administration and on the regulated community. One feature that affected the regulated community was a mandated data requirement. The need for environmental fate data had been recognized by both the CDFA and Environmental Protection Agency (EPA) prior to 1985.[10,11] The kind of data requirement originally envisioned was selective and, in the case of the EPA, based upon a prioritized health risk assessment. However, the Act required a blanket request for all active ingredients contained in pesticides registered for agricultural use. Registrants were required to submit to the CDFA 11 specific study types which were pertinent to judging ground water leaching potential and environmental fate (Table 1). These study types had been drawn chiefly from the EPA's Pesticide Assessment Guidelines.[12,13] These Guidelines describe the kinds and manner of testing required for a selected list of pesticides studies.

The Act was also written with comparatively unambiguous language regarding the scientific methodology such as direct specification of the test types deemed relevant to judging leaching potential and environmental fate. This was in contrast to other environmental legislation which has tended to describe broad goals, while leaving the technical details to the implementing agencies. For example, the California Environmental Quality Act specifies a framework for consideration of various environmental impacts, but leaves the details of assessment methodology to the affected agency.

Valid, Complete, and Adequate

The Act stipulates that the submitted studies "... shall, at a minimum, meet the testing and reporting requirements provided by the Environmental Protection Agency Pesticide Assessment Guidelines..." (Reference 1:Section 13143(b)) as described in Subdivisions D[12] for product chemistry and N[13] for environmental fate. The CDFA is directed to determine if the submitted studies are "... valid, complete, and adequate" (Reference 1:Section 13145(a)). These requirements tied the evaluation process closely to the EPA Guidelines. In most cases the suite of studies for a particular compound were not submitted together; instead, they were submitted at different times. It was not always feasible to determine the potential interaction between two or more processes studied under different test types. This contrasts with other CDFA study review processes which embody a more holistic approach toward an active ingredient and take into account more than just the details of how a test was conducted and how thoroughly it was reported.

The stipulation of valid, complete, and adequate also meant that technical data sheets and study summaries would not suffice, nor would most peer-reviewed scientific articles; it became mandatory to have full study reports with appropriate protocols, analytical results, and quality assurance data, such as calibrations, spike recoveries, and method validation data in some cases. The regulated community did not expect that this level of reporting would be required and, at first, almost nothing submitted met these standards.

In the late 1970s and early 1980s when they were crafted, the EPA Guidelines were, by and large, designed to evaluate singular, moderate molecular weight organic compounds that were applied on the order of pounds per acre. Because the chemistry or use patterns of certain groups of active ingredients did not allow straightforward application of the tests prescribed in the EPA Guidelines, the CDFA deferred the data requirement for such active ingredients. Some examples were preplant seed treatments; inorganic compounds; complex mixtures, such as gasoline and kerosene; plant growth regulators; and naturally occurring materials, such as botanical pesticides. In deferring data requirements for these groups of pesticides, it was the intention of the CDFA to subsequently determine what data would be appropriate to require. Additionally, the CDFA will require only germane data for degradation products and inert ingredients on a selected subset of these compounds or materials.

As of April 1991, more than 6100 individual texts have been presented for evaluation. Total texts include various kinds of submissions which are not actually original test reports, such as rebuttals, waiver requests, and protocols. Original test reports total 4745 (Table 1).

Table 2. CDFA Physicochemical Data Base.
Numbers Shown Are as of March, 1991
for Studies within EPA Guidelines

Property	Records in data base	Number of chemicals
Solubility	321	166
Vapor pressure	301	160
Octanol water	285	156
Hydrolysis	970	174
Water photol.	244	135
Soil degradation		
Aerobic	279	113
Anaerobic	174	99
Photolysis	245	127
Field dissip.	94	44
Soil adsorption		
Kd	748	153
Koc	70	16
Desorption	350	70
Henry's Constant	161	131

Data base

The data requirement, mandated by the Act, provided an opportunity to construct a data base of physicochemical properties for active ingredients. Extensive analysis was devoted to determining appropriate data fields, ranges, and relationships within the intended data base. This analysis was actually started prior to the Act.[10]

The different number of records in each test type in the data base reflects the different requirements of the study types (Table 2). For example, hydrolysis and adsorption must be characterized under a range of conditions. Hydrolysis studies are typically conducted under three different pH regimes. Adsorption studies are typically conducted with at least four different soil types. Consequently, there are more records for these test types. Variability in the number of records also results from intrinsic properties of some compounds which preclude the performance of certain studies. For example, extremely high volatilization may preclude a soil degradation study. The low number of field dissipation studies reflects both fewer required conditions and the difficulty registrants have encountered in conducting this type of study according to the Guidelines. Because of data gaps and the lag time in processing studies, the data base is currently incomplete. In addition, quality control studies are needed to identify and correct errors.

ESTABLISHING THE GROUND WATER PROTECTION LIST (GWPL)

The GWPL, mandated by the PCPA, is used as a list of priority analytes in ground water monitoring programs. An active ingredient is placed on this list

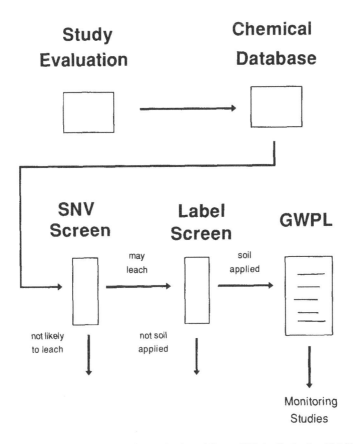

FIGURE 1 Study evaluation and determination of Ground Water Protection List (GWPL) through Specific Numerical Value (SNV) Screen and label screen.

if it passes through a two-staged screening process (Figure 1). The first screening stage is based on physicochemical properties and the second on label language concerning soil application. In the first stage "cutoff points" or specific numerical values (SNVs) are established for certain key physicochemical parameters and used to determine which chemicals have a potential for leaching.[14]

Specific Numerical Values (SNVs)

The Act calls for the development of SNVs based upon water solubility, soil adsorption coefficient (Koc), hydrolysis, aerobic and anaerobic soil metabolism, and field dissipation characteristics. This concept came directly from work being done by the EPA at the time that the PCPA was being drafted. This work was subsequently published by Creeger.[14] The PCPA mandated a test based on two groups of parameters: those associated with mobility (solubility and soil adsorption) and those associated with persistence (hydrolysis, aerobic and anaerobic soil metabolism, and field dissipation). Compounds which were mobile and persistent would be classified as potential leachers. Such compounds would move on to the next stage in the screening process to establish the GWPL.

Since the PCPA did not specify the method for setting the SNVs, the CDFA devised a statistical procedure based on nationwide ground water detection reports.[15-17] This procedure required compiling nationwide ground water monitoring data for agricultural chemicals. Monitoring information about pesticides sampled for in ground water was obtained from published reports. Non-leacher pesticides consisted of those pesticides for which ground water sampling had been conducted, but for which there had not been any positive detections. After compiling this information, a list of leachers and non-leachers was assembled.

Next, the relevant physicochemical parameters were estimated for each listed pesticide. In the first SNV report,[15] sufficient information was available only for solubility, adsorption normalized to organic carbon, hydrolysis, and soil degradation. We examined several criteria for suitability in determining SNVs and decided to use that point for each physicochemical property which captured 90% of the leacher population.[15]

The actual logic of the SNV screen is: classify as a potential leacher if (A or B) and (C or D) is true, where: A = 'Koc less than V1'; B = 'Water solubility (Sol) greater than V2'; C = 'Hydrolysis half-life (Hyd) greater than V3'; D = 'Aerobic soil metabolism half-life (ASM) greater than V4'; and V1, V2, V3, V4 are SNVs.

In the 1989 SNV report,[17] this scheme resulted in SNVs as shown in Figure 2. There was no significant statistical separation between leachers and non-leachers for aerobic soil metabolism in the 1988 revision.[16] Consequently, the half-life was set arbitrarily high in order to effectively remove it from the process, while still satisfying the requirements of the Act. A more recent report, in preparation, includes anaerobic soil metabolism. Field dissipation will be included when sufficient information becomes available.

A majority of the errors in the SNV classification procedure occur because non-leachers are classified as leachers. This occurs because the population distributions overlap even though significant statistical separation occurs between the means (Figure 3). Therefore, a portion of the non-leacher population is included in the region delineated by the SNV as shown in Figure 3 for Koc SNV of 1900 cm^3/g. CDFA is actively investigating alternatives to the SNV procedure to improve the accuracy of classification.

Label Screen

Compounds which have passed through the SNV screen are then evaluated by a second procedure referred to as a label screen. This second procedure examines the label language to determine if any uses on labels registered in California include intentional incorporation or application to the soil by ground-

Specific Numerical
Values Procedure

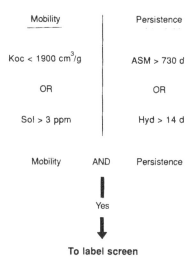

FIGURE 2 Specific numerical values screen applied to compounds as partial test of placement on the Ground Water Protection List. Koc = soil adsorption normalized for carbon, ASM = aerobic soil metabolism half-life, Hyd = hydrolysis half-life, Sol = solubility.

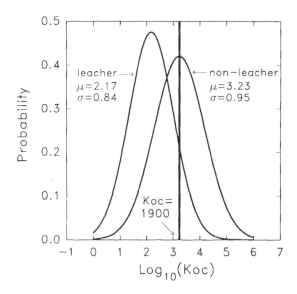

FIGURE 3 Statistical populations for Koc of leachers and non-leachers for 1989 SNV report showing 90th percentile determination for Koc.[17]

based application equipment or chemigation, or application followed by flood or furrow irrigation within 72 h. These label statements have been chosen to distinguish soil-applied pesticides from those which are applied strictly to foliage. If the label language implies any of the soil-applied conditions, then the compound is placed on the GWPL.

Pesticides can also be placed on the GWPL if they have previously been found in ground water in California due to agricultural use and are currently registered. Pesticides on the GWPL receive priority in CDFA ground water monitoring programs. Also the PCPA requires dealers to report all sales of pesticides on the GWPL, and applicators to report all uses to CDFA.

CONCLUSIONS

Executing the mandates of the PCPA involved the creation of the specific numerical values and the publication of the Ground Water Protection List. In order for those tasks to be fully implemented it has been necessary for registrants to generate much chemical and environmental fate data and for CDFA to evaluate it. The on-going tasks to implement the PCPA include the periodic revision of the Ground Water Protection List and the specific numerical values, as well as a continuing program of state-wide ground water monitoring and reevaluation of pesticides detected in ground water.

LITERATURE CITED

1. Pesticide Contamination Prevention Act. Article 15, Chapter 2, Division 7 of the Food and Agricultural Code, Statutes of 1985, Chapter 1298.
2. Hallberg, G. R.; Libra, R. D.; Bettis, E. A., and Hoyer, B. E. *Hydrogeologic and water quality investigations in the Big Spring Basin, Clayton County, Iowa.* Iowa Geological Survey Open File Report 84-4. Iowa City, IA. 1984.
3. Rothschild, E. R.; Manser, R. J.; and Anderson, M. P. *Ground Water* 1982, 20:437–445.
4. Rhodes, H. L. In *Evaluation of Pesticides in Ground Water;* Garner, W. Y., Honeycutt, R. D.; and Nigg, H. H., Eds.; ACS Symposium Series 315; American Chemical Society: Washington, D.C. 1986 pp 541–547.
5. Zaki, M. H.; Moran, D.; and Harris, D. *Am. J. Public Health* 1982, 72(12):1391–1395.
6. Cohen, D. B. In *Evaluation of Pesticides in Ground Water;* Garner, W. Y., Honeycutt, R. D.; and Nigg, H. H., Eds.; ACS Symposium Series 315; American Chemical Society: Washington, D.C. 1986 pp 499–529.
7. Cohen, D. B. and Bowes, G. W. *Water Quality and Pesticides: A California Risk Assessment Program.* State Water Resources Control Board, Toxic Substances Control Program. Sacramento, CA Draft Report 12/7/84.

8. Litwin, Y. J.; Hantzsche, N. N.; and George, N. A. *Groundwater Contamination by Pesticides — A California Assessment*. Ramlit Associates, Inc. Berkeley, CA, 1983, in conjunction with Toxic Substances Control Program, California Water Resources Control Board, Sacramento, CA.

9. Price, P. and Umino, W. (project managers). *The Leaching Fields: A Nonpoint Threat to Groundwater*. Assembly Office of Research, State of California, 1985.

10. Leifson, O. "Data for assessment of environmental fate of pesticides." Memorandum to Lori Johnston. Environmental Monitoring and Pest Management. California Department of Food and Agriculture, December 16, 1983.

11. Anonymous. *Chemical Regulation Reporter* 1984 8(5):139, 149–150.

12. *Pesticide Assessment Guidelines Subdivision D Product Chemistry;* Hitch, R. K., Coord.; EPA-540/9-82-018; U.S. E.P.A.: Washington, D.C., 1982.

13. *Pesticide Assessment Guidelines Subdivision N Chemistry: Environmental Fate;* Hitch, R. K., Coord.; EPA-540/9-82-021; U.S. E.P.A.: Washington, D.C., 1982.

14. Creeger, S. In *Evaluation of Pesticides in Ground Water;* Garner, W. Y., Honeycutt, R. D.; and Nigg, H. H. Eds.; ACS Symposium Series 315; American Chemical Society: Washington, D.C. 1986 pp 548–557.

15. Wilkerson, M. R.; Kim, K. D. *The Pesticide Contamination Prevention Act: Setting Specific Numerical Values*. Environmental Hazards Assessment Program. California Department of Food and Agriculture, State of California. Sacramento, CA, 1986.

16. Johnson, B. *Setting Revised Specific Numerical Values November 1988*. Environmental Hazards Assessment Program. Department of Food and Agriculture, State of California. Sacramento, CA, 1988; EH 88–12.

17. Johnson, B. *Setting Revised Specific Numerical Values October 1989*. Environmental Hazards Assessment Program. Department of Food and Agriculture, State of California. Sacramento, CA, 1989; EH 89–13.

CHAPTER 12

Natural and Man-Made Modes of Entry in Agronomic Areas

James M. DeMartinis and Sandra C. Cooper

Many reported detections of agricultural chemicals in ground water are not the result of leaching through the soil root zone and unsaturated zone under normal agronomic practices. Rather, accidental releases as farm-specific point sources in hydrogeologically vulnerable areas or near man-made conduits have resulted in the direct entry of chemicals to ground water at relatively high concentrations in short periods of time. Chemicals can be introduced to ground water not only via localized spills or runoff, but also by way of direct entry through man-made (faulty well casings; drainage wells) and natural (sinkholes; macropores) pathways. Thus, chemical storage and handling and Best Management Practices are important factors to be considered in preventing point-source releases of agricultural chemicals to ground water.

In the last decade, it has become apparent that pesticide usage can have an impact on ground water quality under certain conditions; however, detections of an agricultural chemical in ground water may or may not be the result of leaching through the soil root zone and the unsaturated zone. While research has shown that under normal agronomic use some more mobile chemicals can leach downward to ground water, it is important to note that this is only one mechanism of movement. In many instances, there are numerous site-specific hydrogeological conditions and/or management practices that control the downward movement of chemicals. In fact, there can be multiple modes of entry responsible for the introduction of agricultural chemicals to ground water.

Mechanisms that can promote the introduction of chemicals to ground water by means other than leaching include natural modes of entry, man-made modes of entry, and site-specific farm management practices. Examples of documented cases where agricultural chemicals have been detected in ground water

0-87371-926-3/94/$0.00+$.50

165

as a result of point source releases include localized spills at land surface; backsiphoning directly into an irrigation well; faulty well construction (the well is not grouted to land surface); agricultural drainage wells (to mitigate runoff problems by directing water into the ground); specific hydrogeologic conditions (fissured clay soils that act as direct conduits to ground water); human errors during sampling and monitoring of chemical movement (cross contamination between samples); and false positive analyses reported by the analytical laboratory.

Field inspections of areas where agricultural chemicals have been reportedly detected in ground water commonly result in nonleaching explanations for the findings. Published information, conversations with local experts and growers, site inspections, and some degree of field work usually are necessary to determine the actual mode of entry by which the chemical reached ground water.

DIRECT ENTRY PATHWAYS

A typical nonleaching scenario usually involves a mode of entry via a direct entry pathway, which can be classified either as man-made (e.g., faulty well seals; drainage wells; improperly abandoned wells) or as natural (e.g., sinkholes; macropores; fissured clay soils). Obviously, it is most important how a particular agricultural chemical is stored, handled, and used in the vicinity of these direct entry pathways. A chemical spill near a direct entry pathway, such as a sinkhole or a faulty well casing, can produce an immediate impact on the ground water quality because the biologically active soil root zone may be bypassed altogether, thus allowing the chemical to reach ground water where residues may not readily degrade.

Direct entry pathways can promote the rapid introduction of an agricultural chemical to ground water within a relatively short period of time after application or accidental spillage rather than months or years which may be required through natural leaching mechanisms. As a result, concentrations of chemical residues entering ground water can be high and potentially reflect higher levels than would normally occur through natural leaching processes. Additionally, once a compound reaches ground water via a direct entry pathway, residues can be dispersed throughout the aquifer (depending on hydrogeologic conditions) and migrate laterally off site.

Natural Modes of Entry

Natural modes of entry refer to natural fissures or cracks or channels that develop in the soil as a result of pedogenic processes or as the result of soil flora and/or fauna activities. These pathways include insect and animal burrows, deeply penetrating plant roots that can create channels, and specific soil properties

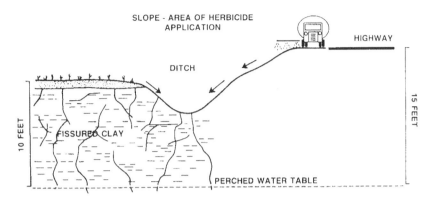

FIGURE 1 Schematic diagram showing fissured soils and relationship to herbicide application.

such as the shrink/swell capacity of a clay. For example, soils containing appreciable quantities of montmorillonitic clay exhibit shrink/swell processes upon alternating wet/dry cycles and can be deeply fractured by cracks after long dry spells. Additionally in karst terranes, development of solution cavities or the dissolution of the underlying limestone can result in the development of sinkholes which can serve as direct conduits to ground water.[1]

An example of a natural mode promoting the introduction of an agricultural chemical to ground water occurred in central California several years ago where an herbicide was sprayed along a steeply sloping roadway to deter the growth of weeds.[2] Applications of the chemical were quite successful, and no one expected to see any impact on ground water quality because the area adjacent to the roadway was underlain by a clay soil. In fact, residues of the herbicide were detected in ground water. The clay soil was highly fissured which promoted the easy transport of the herbicide down the steep slope through the cracks in the clay and directly to the water table (Figure 1). In this particular instance, the depth to water was shallow enough and the cracks in the clay were deep enough to provide direct hydraulic connection between land surface and the water table, thus resulting in residue detections in nearby wells.

Man-Made Pathways

Man-made direct entry pathways refer to structures that provide an artificial means of chemical entry to ground water that would not exist under natural conditions. Examples of these types of man-made structures include corroded or faulty well casings and/or screens; corroded or damaged well pumps and/or pump housings; unprotected well casings (e.g., domestic wells not enclosed in protective shelters); corroded irrigation pipes that can leak discharge water; and poorly constructed wells that are not properly sealed off from the surrounding *in situ* aquifer material and/or not properly sealed off at land surface.

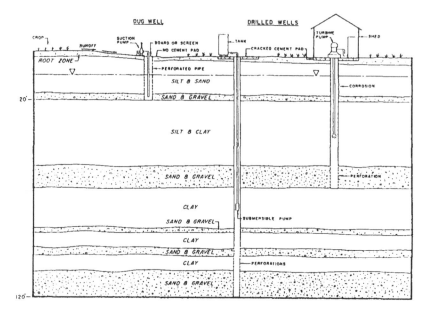

FIGURE 2 Faulty domestic well schematics.

The term "faulty" well is used as a general term for those wells that are not constructed properly, are of dubious integrity, are located in undesirable areas, have been damaged, abandoned, or are not installed to current specifications (Figure 2). In most areas of the country, there are state or local (county, township, or city) guidelines and/or requirements for well specifications. Surface seals to prevent surface water infiltration are common for newly constructed wells; however, many older domestic and irrigation wells that are still in use or that have been abandoned were constructed prior to existing statutes and guidelines.[3]

For example, in one large agricultural area of the country, irrigation wells were constructed to drain excess irrigation water from croplands. The drainage wells (most of which were constructed about 10 to 15 years ago) generally are about 3 ft in diameter, range from 60 to 100 ft in depth, and are packed the entire depth of the annular space with coarse gravel to enhance the well yield (Figure 3). The locations of these irrigation drainage wells are near to or in fields where agricultural chemicals have been applied, or in close proximity to areas where chemicals have been stored, mixed, or handled. The sole purpose of these wells was to accept excess spent irrigation water (containing chemical residues) from the croplands, and, thus, allow water to infiltrate downward to both shallow and deeper water-bearing zones. As a result, chemical residues have impacted not only the water quality of the shallow aquifer, but also the quality of the deeper aquifer, which serves as a source of potable water for the surrounding area.

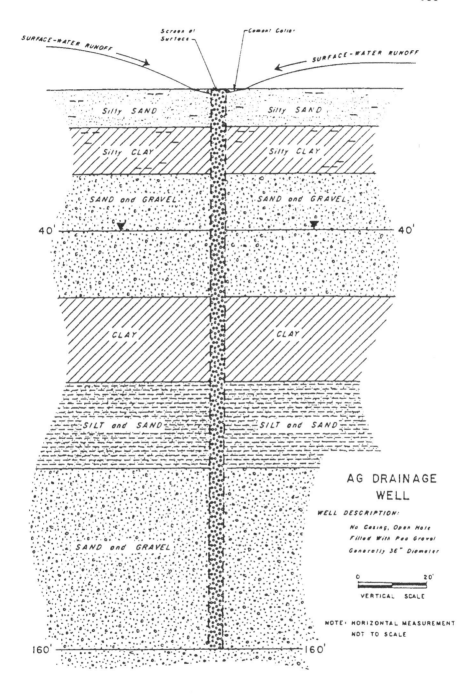

FIGURE 3 AG drainage well.

Newly constructed irrigation wells now are installed with surface casings, and the annular spaces within the wells are now grouted with cement to seal off shallow perched water zones from the deeper water-producing zones, thus preventing spent irrigation water from migrating downward to the deeper aquifers.

Domestic wells also can provide direct entry for chemicals to reach ground water. The authors have observed domestic in-use wells that have no surface seals or badly cracked or damaged surface seals. Other documented observations have shown domestic wells open to the atmosphere, thereby allowing surface debris deposition and direct surface runoff during heavy precipitation to enter the well directly.

Numerous wellheads are sheltered by attractive looking sheds, but offer little if any wellhead protection because the surface seals of the wells are in poor repair. In the case of the homeowner who uses domestic agricultural chemical products, these sheds also provide an attractive dry place for the unwary homeowner to store unused portions of domestic-use pesticides.

The location of domestic wells in close proximity to fields on which agricultural chemicals are applied and/or to chemical mixing/loading areas can present problems for poorly constructed or improperly sheltered wells. Wells situated topographically lower and located near these areas can provide pathways for surface-water runoff (containing chemical residues) to impact ground water quality.[4]

Case History

Blasland, Bouck, & Lee conducted an agricultural chemical study encompassing four counties in the Sacramento Valley in California.[2] The first task included the identification of domestic wells, having confirmed detections of a specific agricultural product as determined through random sampling by a state agency. Data collected for the wells and adjacent areas were used to determine whether a correlation existed among well characteristics, local agronomic practices, soil types, and residue detections. Tables 1 and 2 present summaries of the well data collected through personal interviews, telephone conversations with well owners, and visual inspections.

The findings in Table 2 demonstrate that upon close inspection of the sampled wells, it was determined that a number of the wells themselves may have provided direct entry pathways for agricultural chemicals to enter ground water. Additionally, it was learned that these same well owners saw no problems with the construction or protection of their wells despite the fact that an average of 20% of the randomly selected wells in all four counties were open to the atmosphere. Of the total number of wells inspected, an average of 41% had either poor or no surface seals and an average of 4% had corroded/cracked casings. Additionally, some wells (an average of about 14%) were topographically lower and as close as 20 ft in proximity to chemical application areas. Based on these findings, the probability of these wells acting as direct entry

Table 1. Well Data Summary for Four Counties

Well data information	County			
	W	X	Y	Z
Average well depth[a]	130	140	110	90
Type construction[b]				
Drilled	15	5	7	6
Dug	1	0	0	1
Driven	0	0	0	0
Date installed[b]				
Before 1920	3	0	1	2
1921–1950	2	0	1	1
1951–1980	10	4	5	3
1981–present	1	1	0	1
Average distance to chemically treated area[c]	330	190	230	310
Average depth to ground water[a]	13	15	18	11

[a] Well depths and depths to water are in feet below land surface.
[b] Well construction types and dates are for a total of 35 wells in the four-county area. These 35 appear to be representative of all wells visited. However, complete data were not available for the approximate other 60 wells evaluated. Particularly, the well depth and date of installation were difficult to obtain or verify. The reader is referred to in Reference 11 regarding the availability of domestic well records in the U.S.
[c] Distances to treated areas are in feet.

Table 2. Statistical Summary of Well Data for Four Counties

Well data information[a]	County				
	W	X	Y	Z	AVG
Poor surface seals	17	0	57	33	27
No surface seals	8	0	14	33	14
Cracked/corroded casings	17	0	0	0	4
Topographically lower than adjacent fields	25	0	14	17	14
Protected by shed/well house	50	33	14	33	33
Chemical storage in shed	17	20	0	0	9
Well head open to atmosphere	17	0	28	33	20
Close proximity to storage or mixing/loading area	33	67	28	0	32
Unable to inspect	25	33	0	14	18

[a] Well data information are given as the percentage of wells in each county and the average percentage for the four counties. Includes the 35 wells referred to in Table 1 and others in the region that were visited (approximately 60 total wells).

pathways for agricultural chemicals to reach ground water was determined to be high.

In this particular study area, oil and gas exploration maps also were examined. In a 6-square-mile area where seven domestic wells showed detections of agricultural chemical residues, 23 abandoned oil and/or gas wells and seven wells still producing were identified. In the field, no surface expression of the abandoned wells were observed visually, and, for the most part, these abandoned wells were located in cropped fields under cultivation. Thus, the potential existed for these abandoned wells to provide direct entry pathways to ground water.

As part of this same study, agricultural drainage wells in the area also were examined. In the Central Valley of California, approximately 5000 abandoned drainage wells were capable of channeling surface runoff directly to ground water.[5] Drainage wells also have been noted in other areas of the country, and particularly in north central Iowa.[6] Identification of agricultural drainage wells is critical, as they can serve as direct entry pathways for chemical residues to reach ground water rapidly, thus negating natural degradation processes. Although drainage wells were not found in the immediate vicinity of the study area, their presence was confirmed south of the study area in the San Joaquin Valley.[2]

This case history demonstrates the need to examine closely both farm and domestic well construction/location/protection conditions as providing potential direct entry pathways to ground water for agricultural chemicals. Additionally, irrigation wells, abandoned wells, oil and gas wells, and agricultural drainage wells on or near agricultural areas must be considered suspect when investigating reported detections of chemical residues in ground water.

PREVENTION OF FARM-SPECIFIC POINT SOURCE RELEASES

Because man-made structures such as wells can provide direct conduits for the migration of agricultural chemicals to ground water, the resulting impact on water quality can be rapid. Once chemical residues reach ground water, remediation often is difficult and costly. The following is a brief, albeit not exhaustive, description of simple preventative measures that can be used to minimize point source releases of agricultural chemicals. The reader is referred to References 7 and 8 for further information.

Well Inspections

All well (farm, domestic, irrigation, drainage, etc.) should be inspected periodically for cracked casings and/or damaged well seals. A tight cement or bentonite seal along the annular space surrounding the well casing will minimize the potential for surface runoff to infiltrate down along the outside of the casing. Raising the area around the well will prevent drainage from flowing towards the well. Additionally, a tight seal will prevent ground water from a shallow degraded zone from migrating downward to deeper unimpacted water-bearing units. In general, clay-based grouts such as bentonite provide a low-permeability seal that remains flexible over time. Bentonite tends to self-seal if the well casing (or even the land surface) should shift over time, thus lasting longer and acting as a more effective seal than cement.

Spray Equipment

Spray equipment used in the application of agricultural chemicals should be cleaned and rinsed in designated areas and as far away from any wells as possible. Ideally, the clean/rinse area should be on topographically lower ground than the well head to minimize the potential for runoff to the well. Obviously, a rinsate pad constructed of cement and mounded along the edges is preferable to contain rinsate flushed from the spray equipment. Finally, proper disposal of rinsate (according to label instructions) is necessary.

Empty Product Containers

Empty or partially empty product containers should not be stored temporarily near any wells and should be disposed of properly. Improper disposal of used containers, such as dumping in topographically low areas or even in sinkholes, unfortunately, has been an all too common practice in some areas.

Chemigation

Agricultural chemicals can be introduced directly into a well by chemigation, which is the practice of applying chemicals through use of an irrigation system. If the irrigation pump fails, chemical injection may continue resulting in chemical backflow down the well.[9] An ineffective or inoperative backflow prevention device on the water supply line can cause the water-chemical mixture to flow back into the well. Therefore, irrigation and pumping equipment should be monitored and maintained.

Chemical Storage

Storage of agricultural chemicals away from wells is essential. Storage of chemicals in the well house or shelter also is unacceptable. Likewise, storage near abandoned wells or sinkholes should be avoided.

Chemical Mixing

Mixing and loading agricultural chemicals should not take place near wells. Many "catastrophic" and longer term spills have been documented in the literature, and the negative effects on ground water quality have been measured.[10,11] However, very little attention has been focused on the farm level. During the mixing and loading process, tank overflow can occur and backsiphoning can cause agricultural chemicals to enter wells directly.

Newly Constructed Wells

Installing new wells in topographically high areas above floodplains of streams and away from all potential sources of contamination (e.g., leachfields, feedlots, mixing/loading areas, etc.) is essential. Acceptable "setback" distances for the location of wells depends on soil types and ground water vulnerability criteria, but generally are mandated by local health departments. The EPA has well-established well head protection guidelines.[12]

Decommissioning Old Wells

All abandoned and unused wells should be sealed according to legal specifications to eliminate the potential for contaminated surface runoff to enter the wells. Another equally important reason for properly decommissioning old wells is to eliminate the potential for shallow degraded ground water to mix with "clean" ground water in deeper water-bearing zones. Field observations have been documented where unsealed abandoned wells on farms were allowed to deteriorate to the point that the wells became direct entry pathways.[3,4]

Protection of Vulnerable Hydrogeologic Areas

Sinkholes and other natural features (e.g., macropores, animal burrows, solution cavities) can be protected from surface-water runoff by grass filter strips which act as buffer zones. In areas where the slope of the land surface is steep, a more sophisticated rerouting system for drainage water may be required. Sealing sinkholes or macropores has been attempted but is generally not recommended because of the difficulty in obtaining tight effective seals. Keeping agricultural chemicals away from all such features has generally been the recommended course of action.[7]

CONCLUSIONS

There are multiple modes of entry for agricultural chemicals to reach ground water, and leaching is only one mechanism. There are both natural and man-made modes of entry in which chemicals can be introduced rapidly and directly into the ground water system with the outcome resulting in high residue concentrations of the chemical. Additionally, management practices most certainly can contribute to the problem of point source releases of agricultural chemicals.

Identification of farm-specific direct entry pathways and prevention of point source releases of chemicals is critical in minimizing local ground water problems and preventing regional impacts on ground water quality. Once an agricultural chemical reaches ground water, it may not degrade easily, and the resulting monitoring and remediation becomes both difficult and costly. Preventing ground water impacts from agricultural chemicals by following proper handling/loading/mixing procedures; by identifying potential direct entry pathways; and by using Best Management Practices is essential on both the farm and homeowner level.

LITERATURE CITED

1. Libra, R. D., Hallberg, G. R. and Hoyer, B. E., 1987; Impacts of Agricultural Chemicals on Ground Water Quality in Iowa, in *Ground Water Quality and Agricultural Practices,* Lewis Publishers, Chelsea, Michigan, pp. 185–216.
2. DeMartinis, J. M. and Royce, K. L., 1990; Identification of Direct-Entry Pathways by which Agricultural Chemicals Enter Ground Water, Proceedings of National Water Well Association Cluster of Conferences; pp. 51–66.
3. Blomquist, P. K., 1984; Abandoned Water Wells in Southern Minnesota, Proceedings from the Seventh National Ground Water Quality Symposium, NWWA, p. 330–344.
4. Exner, M. E. and Spalding, R. F., 1985, Ground-Water Contamination and Well Construction in Southeast Nebraska, *Ground Water,* Vol. 23, No. 1, pp. 26–34.
5. Holden, P. W., 1986; *Pesticides and Ground Water Quality, Issues and Problems in Four States;* National Academy Press, Washington D.C.
6. Drake, L. and Esling, S., 1988; Regional Hydrogeology of Agricultural Drainage Wells in Iowa, *Geological Society of America Bulletin,* Vol. 100; No. 7, pp. 1120–1130.
7. Office of Technology Assessment, 1990; Beneath the Bottom Line: Agricultural Approaches to reduce Agrichemical Contamination of Ground Water; Congress of the United States, OTA F-418, Washington, D.C.
8. University of Wisconsin-Extension and Wisconsin Department of Agriculture, Trade and Consumer Protection, 1989; Nutrient and Best Management Practices for Wisconsin Farms, Wisconsin Department of Agriculture, Trade and Consumer Protection Technical Bulletin ARM-1, 174 p.
9. Schepers, J. S. and Hay, D. R., 1987; Impacts of Chemigation on Ground Water Contamination, in *Ground Water Contamination,* Lewis Publishers, Chelsea, Michigan, pp. 105–114.
10. Ryckman, M. D., 1984; Detoxification of Soils, Water and Burn Residues from a Major Agricultural Chemical Warehouse Fire, The 5th National Conference on Management of Uncontrolled Hazardous Waste Sites, November 1984, Washington, D.C., pp. 420–426.
11. Sterrett, R. J., Barnhill, G. D., and Ransom, M. E., 1985; Site Assessment and On-Site Treatment of a Pesticide Spill in the Vadose Zone, NWWA Conference on the Vadose Zone, Nov. 19–21, 1985, pp. 255–271.
12. USEPA, 1987, Guidelines for Delineation of Wellhead Protection Areas, Office of Ground-Water Protection, EPA 440/6-87-010.
13. Ganley, M. C., 1989; Availability and Content of Domestic Well Records in the United States, *Ground Water Monitoring Review,* Vol. 9, No. 4, pp. 149–158.

CHAPTER 13

Drinking Water — A Farm Woman's Perspective

Sandra Hayes Greiner

As a farmer, mother, and consumer, I truly consider myself to be "caught in the middle" in my previous life. I was a "traditional" farm wife — now, I am an extremely nontraditional farm wife. I'm probably uniquely qualified to address this subject because of the assortment of responsibilities that I assume throughout the year.

My husband, Terry, is a seed corn dealer, as well as a pork producer and row crop farmer. As his spouse and partner, I come in contact with a variety of farmers with various levels of sophistication throughout the growing season.

I spend an entire week during "Pioneer Days" visiting with farmers as they wait in our office to meet with my husband to order their seed.

I play bridge with the farmers' wives. I sit with them on the bleachers during little league baseball games and high school basketball games. I work in concession stands sponsored by band mothers, serve on various rural economic development committees, and we sing in the church choir together. We work in political organizations, working together to elect candidates who understand the challenges we face as farmers to grow food for a hungry world.

During the spring planting season, which is basically April and May, I run our seed warehouse while my husband plants our crops. It is this responsibility that provides me the opportunity to interact with the farmers on a more or less equal basis.

It is this experience, running the forklift, keeping track of who needs seed, where and when, and visiting with the farmers on a one-on-one basis as they

0-87371-926-3/94/$0.00+$.50

pick up corn or soybeans, that I have come to the conclusion that their various levels of sophistication really make no difference in how they view crop protectors.

Whether my seed customers grow crops on 80 acres or 880 acres seems to make no difference in their attitude towards crop protectors. It doesn't seem to matter whether the customer has an agronomy degree from Iowa State, or a high school diploma from good ole KHS. None of these factors seem to fit into a pattern that would indicate that one group or another is concerned about possible toxicity. The one clear message that I have noticed coming from this cross section of German Catholic farmers is that they consider themselves to be careful and conscientious stewards of our natural resources. They consider their neighbors to be responsible stewards as well. Their concern is for the fellows that live two hundred miles north of them ... that they might not be as careful as the farmers in our community. It is, quite frankly, the fear of the unknown.

Having given you the background on the farmers that I know personally and deal with on a yearly basis, I can tell you unequivocally that none of these farmers, for economic reasons, likes the idea of using crop protectors ... neither do they like foxtail, black nightshade, spider mites, or corn bores. These same farmers do not like the idea of the high cost of medical insurance premiums, nor the risk of attempting to insure themselves.

Currently, there is a big trend in my neighborhood to move to no-till soybeans. This, of course, will help us comply with our plans filed at the local ASCS office to reduce soil erosion on our farms. Yet, drilling seed into an undisturbed seed bed has a real potential for some serious weed problems. It seems to me that we can't have our cake and eat it too.

It has occurred to me that you put yourselves at considerable risk each planting season. As farmers hurry to the field and begin their seed bed preparation, racing against the weather, they apply crop protection chemicals under the stress of trying to grow a little better crop with a little less input.

Obviously, we are dealing with a two-headed monster ... as a producer, we must trust that you have researched the capabilities of these various products and labelled them properly in order that we apply them at the proper calibration rates, but, on the other hand, you must trust that every one of those farmers is conscientious enough to be sure that they are applying the product at the recommended application rate. Your integrity depends on a bunch of farmers racing against the clock, the calendar, and Mother Nature. Yet, I can assure you that I know of no farmer who would intentionally do anything to risk his soil fertility. From a personal perspective, Terry and I decided early on to put all of our financial resources back into the farm. The greatest legacy Terry and Sandy Greiner can leave to our children is love and respect for the natural resources ... in our case, good ole Iowa dirt.

As a consumer, I prefer the quality, abundance, and affordability that crop protectors provide. Last March I was invited to speak to my Oregon affiliates annual meeting. Part of their program was a panel of organic farmers. After the presentation, during the question and answer session, one of the ladies attending the meeting told about attending an organic seminar somewhere in Oregon in order to investigate the possibility of shifting her commercial apple orchard to organic. She told about visiting with a woman at this seminar and asking about how she treated some particular production problem. The grower went into great detail explaining that she lived across the road from a poultry farm and that the poultry producer was forced to pay someone to come in to remove poultry droppings twice a week and how expensive this was to the poultry farmer. The apple producer had come up with the possibility of taking the droppings and using them in her apple orchard. To make a long story short, she was collecting the chicken manure, mixing it with water in a big vat until it became a thin gruel, and spraying it on her apples in an attempt to handle this production problem. It was at least two months before I had the appetite to eat an apple.

As a mother of three sons who constantly trailed along behind their dad from the time they were potty-trained until the time they were finally allowed to run a tractor, I admit to having had some misgivings about the effects on my children from exposure to agricultural chemicals.

Quite honestly, my biggest concern has been the fear that I would prove to be the weak link in the farm safety chain. Am I handling the work clothes properly during laundering? Have I done an adequate job of educating my sons about the proper handling methods since they aren't required to have an applicators' license?

When I'm not holding fort in the seed warehouse, hauling meals to the field, doing bookwork and record keeping, or going for parts, I am the president of American Agri Women, a coalition of farm, ranch, and agri business women. My presidency finds me traveling around the country visiting my affiliates or addressing an assortment of producer and consumer groups concerning animal rights, food safety, or conserving our natural resources.

From this wide experience, I can assure you that farmers, ranchers, and growers are worried about the influence of environmental extremist groups. It sometimes seems that very few people are concerned about where their next meal is coming from.

A large percentage of farmers, ranchers, and growers get their drinking water from private wells, or rural water districts which are supplied by ground water. We share the concern of those who claim that this water is contaminated by foreign matter — be it feedlot runoff, industrial waste, or field runoff. Our first concern is for the safety of our family and friends who drink this water daily, as well as total strangers that our farming operations might affect. However, I don't think it's being unnecessarily arbitrary that we expect proof that we are responsible for that contamination after we are accused, instead of somebody charging that our cropping methods are endangering our natural resources, legislating change, and then never completing the studies to verify that the risk is indeed real.

In an evaluation of the effectiveness of information dissemination in the 1987 integrated farm management demonstration project, researchers found a majority of farmers felt that agricultural chemicals used properly were not endangering the environment, and an even larger percentage agreed on the need for continued use of crop protection chemicals.

A 1990 evaluation of field demonstrations resulted in an interesting chart. Where I come from, all of these reasons are economic. A *Cedar Rapids Gazette* article in December, 1990, reported on a survey of over 800 farmers in 42 states concerning the use of agricultural chemicals.

Let's have a closer look at this chart. I've always said, you can almost tell a person's vintage by the size of their visuals.

In short, there's a lot of paranoia out there. Farmers and ranchers truly believe that our city cousins think that we sit up nights developing elaborate schemes to poison them.

I have a personal opinion as to why farmers in Iowa have developed this inferiority complex, but I can't offer an explanation as to why farmers in the other 41 states where Jeff Davis conducted interviews feel the way they do. We've come a long way in the last 25 years. As a bride, I can remember taking meals to the field in the springtime to a husband who wore no protective clothing, no gloves, no goggles ... and he wouldn't take time to wash his hands before he ate his lunch. Thanks to outstanding educational presentations by manufacturers, producer meetings sponsored by the extension service, and a lot of vicious nagging on my part, he now washes his hands before he eats.

In closing, I should point out that farmers everywhere, regardless of the crop they grow, are constantly monitoring their cash flow. From time to time we see "studies" that indicate farmers want to discontinue the use of agricultural chemicals, but very few operators will do so until their neighbor across the fence does.

Farmers take great pride in the way their crops look. Straight rows, weedless fields, and a healthy looking crop are a tremendous source of pride and accomplishment.

I am excited about the prospects of crop protectors that aren't toxic to humans, animals, and friendly flora. I know that my counterparts across the continent agree. But I can assure you that as long as it's profitable and safe to use crop protection chemicals, American farmers will continue to use them.

On behalf of America's farmers and the American consumer, I thank you for the vital part you play in helping us to produce the safest, least expensive, and most abundant food supply in the history of the world.

INDEX